# 182 Topics in Current Chemistry

Springer-Verlag Berlin Heidelberg GmbH

# Density Functional Theory III

## Interpretation, Atoms, Molecules and Clusters

Volume Editor: R. F. Nalewajski

With contributions by
J. A. Alonso, L. C. Balbás, A. Berces,
R. O. Jones, V. Sahni, T. Ziegler

With 38 Figures and 27 Tables

 Springer

This series presents critical reviews of the present position and future trends in modern chemical research. It is addressed to all research and industrial chemists who wish to keep abreast of advances in the topics covered.

As a rule, contributions are specially commissioned. The editors and publishers will, however, always be pleased to receive suggestions and supplementary information. Papers are accepted for "Topics in Current Chemistry" in English.

In references Topics in Current Chemistry is abbreviated Top.Curr.Chem. and is cited as a journal.

Springer WWW home page: http://www.springer.de

ISSN 0340-1022
ISBN 978-3-662-14839-6      ISBN 978-3-540-49952-7 (eBook)
DOI 10.1007/978-3-540-49952-7

Library of Congress Catalog Card Number 74-644622

© Springer-Verlag Berlin Heidelberg 1996

Originally published by Springer-Verlag Berlin Heidelberg New York in 1996.
Softcover reprint of the hardcover 1st edition 1996

Typesetting: Macmillan India Ltd., Bangalore-25
SPIN: 10536998      66/3020 – 5 4 3 2 1 0 – Printed on acid-free paper

# Volume Editor

Prof. R. F. Nalewajski
Jagiellonian University
Faculty of Chemistry
ul. Ingardena 3
30-060 Krakow, Poland

# Editorial Board

# Foreword

Density functional theory (DFT) is an entrancing subject. It is entrancing to chemists and physicists alike, and it is entrancing for those who like to work on mathematical physical aspects of problems, for those who relish computing observable properties from theory and for those who most enjoy developing correct qualitative descriptions of phenomena in the service of the broader scientific community.

DFT brings all these people together, and DFT needs all of these people, because it is an immature subject, with much research yet to be done. And yet, it has already proved itself to be highly useful both for the calculation of molecular electronic ground states and for the qualitative description of molecular behavior. It is already competitive with the best conventional methods, and it is particularly promising in the applications of quantum chemistry to problems in molecular biology which are just now beginning. This is in spite of the lack of complete development of DFT itself. In the basic researches in DFT that must go on, there are a multitude of problems to be solved, and several different points of view to find full expression.

Thousands of papers on DFT have been published, but most of them will become out of date in the future. Even collections of works such as those in the present volumes, presentations by masters, will soon be of mainly historic interest. Such collections are all the more important, however, when a subject is changing so fast as DFT is. Ative workers need the discipline imposed on them by being exposed to the works of each other. New workers can lean heavily on these sources to learn the different viewpoints and the new discoveries. They help allay the difficulties associated with the fact that the literature is in both physics journals and chemistry journals. [For the first two-thirds of my own scientific career, for example, I felt confident that I would miss nothing important if I very closely followed the Journal of Chemical Physics. Most physicists, I would guess, never felt the need to consult JCP. What inorganic or organic chemist in the old days took the time to browse in the physics journals?] The literature of DFT is half-divided, and DFT applications are ramping into chemical and physical journals, pure and applied. Watch JCP, Physical Review A and Physical Review B, and watch even Physical Review Letters, if you are a chemist interested in applying DFT. Or ponder the edited volumes, including the present two. Then you will not be surprised by the next round of improvements in DFT methods. Improvements are coming.

The applications of quantum mechanics to molecular electronic structure may be regarded as beginning with Pauling's Nature of the Chemical Bond, simple molecular orbital ideas, and the Huckel and Extended Huckel Methods. The molecular orbital method then was systematically quantified in the Hartree-Fock SCF Method; at about the same time, its appropriateness for chemical description reached its most elegant manifestation in the analysis by Charles Coulson of the Huckel method. Chemists interested in structure learned and taught the nature of the Hartree-Fock orbital description and the importance of electron correlation in it. The Hartree-Fock single determinant is only an approximation. Configurations must be mixed to achieve high accuracy. Finally, sophisticated computational programs were developed by the professional theoreticians that enabled one to compute anything. Some good methods involve empirical elements, some do not, but the road ahead to higher and higher accuracy seemed clear: Hartree-Fock plus correction for electron correlation. Simple concepts in the everyday language of non theoretical chemists can be analyzed (and of course have been much analyzed) in this context.

Then, however, something new came along, density functional theory. This is, of course, what the present volumes are about. DFT involves a profound change in the theory. We do not have merely a new computational gimmick that improves accuracy of calculation. We have rather a big shift of emphasis. The basic variable is the electron density, not the many-body wavefunction. The single determinant of interest is the single determinant that is the exact wavefuntion for a noninteracting (electron-electron repulsion-less) system corresponding to our particular system of interest, and has the same electron density as our system of interest. This single determinant, called the Kohn-Sham single determinant, replaces the Hartree-Fock determinant as the wavefunction of paramount interest, with electron correlation now playing a lesser role than before. It affects the potential which occurs in the equation which determines the Kohn-Sahm orbitals, bus once that potential is determined, there is no configurational mixing or the like required to determine the accurate electron density and the accurate total electronic energy. Hartree-Fock orbitals and Kohn-Sham orbitals are quantitatively very similar, it has turned out. Of the two determinants, the one of Kohn-Sham orbitals is mathematically more simple than the one of Hartree-Fock orbitals. Thus, each KS orbital has its own characteristic asymptotic decay; HF orbitals all share in the same asymptotic decay. The highest KS eigenvalue is the exact first ionization potential; the highest HF eigenvalue is an approximation to the first ionization potential. The KS effective potential is a local multiplicative potential; the HF potential is nonlocal and nonmultiplicative. And so on. When at the Krakow meeting I mentioned to a physicist that I thought that chemists and physicists all should be urged to adopt the KS determinant as the basic descriptor for electronic structure, he quickly replied that the physicists had already done so. So, I now offer that suggestion to the chemistry community.

On the conceptual side, the powers of DFT have been shown to be considerable. Without going into detail, I mention only that the Coulson work referred to above anticipated in large part the formal manner in which DFT describes molecular changes, and that the ideas of electronegativity and hardness fall into place, as do Ralph Pearson's HSAB and Maximum Hardness Principles.

It was Mel Levy, I think who first called density functional theory a charming subject. Charming it certainly is to me. Charming it should be revealed to you as you read the diverse papers in these volumes.

Chapel Hill, 1996                                                              Robert G. Parr

# Foreword

Thirty years after Hohenberg and myself realized the simple but important fact that the theory of electronic structure of matter can be rigorously based on the electronic density distribution $n(r)$ a most lively conference was convened by Professor R. Nalewajski and his colleagues at the Jagiellonian University in Poland's historic capital city, Krakow. The present series of volumes is an outgrowth of this conference.

Significantly, attendees were about equally divided between theoretical physicists and chemists. Ten years earlier such a meeting would not have had much response from the chemical community, most of whom, I believe, deep down still felt that density functional theory (DFT) was a kind of mirage. Firmly rooted in a tradition based on Hartree Fock wavefunctions and their refinements, many regarded the notion that the many electron function, $\Psi(r_1 \ldots r_N)$ could, so to speak, be traded in for the density $n(r)$, as some kind of not very serious slight-of-hand. However, by the time of this meeting, an attitudinal transformation had taken place and both chemists and physicists, while clearly reflecting their different upbringings, had picked up DFT as both a fruitful viewpoint and a practical method of calculation, and had done all kinds of wonderful things with it.

When I was a young man, Eugene Wigner once said to me that understanding in science requires understanding from *several different points of view*. DFT brings such a new point of view to the table, to wit that, in the ground state of a chemical or physical system, the electrons may be regarded as a *fluid* which is fully characterized by its density distribution, $n(r)$. I would like to think that this viewpoint has enriched the theory of electronic structure, including (via potential energy surfaces) molecular structure; the chemical bond; nuclear vibrations; and chemical reactions.

The original emphasis on electronic ground states of non-magnetic systems has evolved in many different directions, such as thermal ensembles, magnetic systems, time-dependent phenomena, excited states, and superconductivity. While the abstract underpinning is exact, implementation is necessarily approximate. As this conference clearly demonstrated, the field is vigorously evolving in many directions: rigorous sum rules and scaling laws; better understanding and description of correlation effects; better understanding of chemical principles and phenomena in terms of $n(r)$; application to systems consisting of thousands of atoms; long range polarization energies; excited states.

Here is my personal wish list for the next decade: (1) An improvement of the accurary of the exchange-correlation energy $E_{xc}[n(r)]$ by a factor of 3-5. (2) A practical, systematic scheme which, starting from the popular local density approach, can – with sufficient effort – yield electronic energies with any specified accuracy. (3) A sound DFT of excited states with an accuracy and practicality comparable to present DFT for ground states. (4) A practical scheme for calculating electronic properties of systems of $10^3$ - $10^5$ atoms with "chemical accuracy". The great progress of the last several years made by many individuals, as mirrored in these volumes, makes me an optimist.

Santa Barbara, 1996                                                               Walter Kohn

# Preface

Density functional methods emerged in the early days of quantum mechanics; however, the foundations of the modern density functional theory (DFT) were established in the mid 1960 with the classical papers by Hohenberg and Kohn (1964) and Kohn and Sham (1965). Since then impressive progress in extending both the theory formalism and basic principles, as well as in developing the DFT computer software has been reported. At the same time, a substantial insight into the theory structure and a deeper understanding of reasons for its successes and limitations has been reached. The recent advances, including new approaches to the classical Kohn-Sham problem and constructions of more reliable functionals, have made the ground-state DFT investigations feasible also for very large molecular and solid-state systems (of the order of $10^3$ atoms), for which conventional CI calculations of comparable accuracy are still prohibitively expensive. The DFT is not free from difficulties and controversies but these are typical in a case of a healthy, robust discipline, still in a stage of fast development. The growing number of monographs devoted to this novel treatment of the quantum mechanical many body problem is an additional measure of its vigor, good health and the growing interest it has attracted.

In addition to a traditional, solid-state domain of appplications, the density functional approach also has great appeal to chemists due to both computational and conceptual reasons. The theory has already become an important tool within quantum chemistry, with the modern density functionals allowing one to tackle problems involving large molecular systems of great interest to experimental chemists. This great computational potential of DFT is matched by its already demonstrated capacity to both rationalize and quantify basic classical ideas and rules of chemistry, e.g., the electronegativity and hardness/softness characteristics of the molecular electron distribution, bringing about a deeper understanding of the nature of the chemical bond and various reactivity preferences. The DFT description also effects progress in the theory of chemical reactivity and catalysis, by offering a "thermodynamic-like" perspective on the electron cloud reorganization due to the reactant/catalyst presence at various intermediate stages of a reaction, e.g. allowing one to examine the relative importance of the polarization and charge transfer components in the resultant reaction mechanism, to study the influence of the infinite surface reminder of cluster models of heterogeneous catalytic systems, etc.

The 30th anniversary of the modern DFT was celebrated in June 1994 in Cracow, where about two hundred scientists gathered at the ancient Jagiellonian University Robert G. Parr were the honorary chairmen of the conference. Most of the reviewers of these four volumes include the plenary lecturers of this symposium; other leading contributors to the field, physicists and chemists, were also invited to take part in this DFT survey. The fifteen chapters of this DFT series cover both the basic theory (Parts I, II, and the first article of Part III), applications to atoms, molecules and clusters (Part III), as well as the chemical reactions and the DFT rooted theory of chemical reactivity (Part IV). This arrangement has emerged as a compromise between the volume size limitations and the requirements of the maximum thematic unity of each part.

In these four DFT volumes of the *Topics in Current Chemistry* series, a real effort has been made to combine the authoritative reviews by both chemists and physicists, to keep in touch with a wider spectrum of current developments. The Editor deeply appreciates a fruitful collaboration with Dr. R. Stumpe, Dr. M. Hertel and Ms B. Kollmar-Thoni of the Springer-Verlag Heidelberg Office, and the very considerable labour of the Authors in preparing these interesting and informative monographic chapters.

Cracow, 1996                                                                    Roman F. Nalewajski

# Table of Contents

# Table of Contents of Volume 180

# Table of Contents of Volume 181

# Table of Contents of Volume 183

## Density Functional Theory IV: Theory of Chemical Reactivity

# Quantum-Mechanical Interpretation of Density Functional Theory

Viraht Sahni

Department of Physics, Brooklyn College of the City University of New York, 2900 Bedford
Avenue, Brooklyn, New York 11210,
and
The Graduate School and University Center of the City University of New York, 33 West 42nd
Street, New York, New York 10036

## Table of Contents

Topics in Current Chemistry, Vol. 182
© Springer-Verlag Berlin Heidelberg 1996

This article describes the rigorous quantum-mechanical interpretation of Hohenberg–Kohn–Sham density-functional theory based on the original ideas of Harbola and Sahni, and of their extension by Holas and March. The *local* electron-interaction potential $v_{ee}^{KS}(r)$ of density-functional theory is defined *mathematically* as the functional derivative $v_{ee}^{KS}(r) = \delta E_{ee}^{KS}[\rho]/\delta\rho(r)$, where $E_{ee}^{KS}[\rho]$ is the electron-interaction energy functional of the density $\rho(r)$. This functional and its derivative incorporate the effects of Pauli and Coulomb correlations as well as those of the correlation contribution to the kinetic energy. The potential $v_{ee}^{KS}(r)$ also has the following *physical* interpretation. It is the work done to move an electron in a field $\mathscr{F}(r)$, which is the sum of two fields. The first, $\mathscr{E}_{ee}(r)$, is representative of Pauli and Coulomb correlations, and is determined by Coulomb's law from its source charge which is the pair-correlation density. The second field, $Z_{t_c}(r)$, represents the correlation contribution to the kinetic energy, and is proportional to the difference of fields derived from the kinetic-energy-density tensor for the interacting and non-interacting systems. The field $\mathscr{F}(r)$ is conservative, and thus the work done in this field is path-independent. The quantum-mechanical electron-interaction energy component $E_{ee}[\rho]$ of $E_{ee}^{KS}[\rho]$ is the energy of interaction between the electronic and pair-correlation densities. The correlation-kinetic-energy component $T_c[\rho]$ can also be written in terms of its source through the field $Z_{t_c}(r)$. Some results for finite atomic (both ground and excited states) and extended metal surface systems derived via the interpretation are presented. Certain consequences of the physical interpretation such as the understanding of Slater theory, and the implications with regard to electron correlations within approximate Kohn–Sham theory are also discussed.

# 1 Introduction

In Schrödinger theory [1] the correlations between electrons are incorporated in the structure of the stationary state wavefunction $\Psi$ of the system. These correlations arise due to the Pauli exclusion principle and Coulomb repulsion, the former being accounted for by the requirement that the wavefunction be antisymmetric in an interchange of the co-ordinates (including spin) of any two electrons. Due, however, to the two-particle electron-interaction operator in the Hamiltonian, the analytical dependence of the wavefunction on the electronic coordinates representative of Coulomb correlations is unknown. Properties of the system are determined as expectation values, taken with respect to the wavefunction, of operators representing the observables of interest. Thus the energy is the expectation value of the Hamiltonian. Now according to the first theorem of modern density-functional theory [2] due to Hohenberg and Kohn [3], the ground-state wavefunction $\Psi$ is a functional of the exact ground-state electronic density $\rho(\mathbf{r})$. Thus the ground-state expectation value of any observable, and therefore the energy, is a unique functional of the density. The ground-state density, however, does not discriminate between interacting and noninteracting electronic systems. Thus, in the Kohn–Sham [4] version of density-functional theory, the Schrödinger (Kohn–Sham) equation for a model system of *noninteracting* quasi-particles which leads to the same density is solved instead. Since these quasi-particles are noninteracting, the operator (potential) representing all the electron correlations, including those of the correlation contribution to the kinetic energy, is *local*. Furthermore, as a consequence of the second theorem of Hohenberg and Kohn [3] which establishes the variational character of the ground-state energy functional, Kohn–Sham theory provides a *rigorous mathematical* definition for this local potential. It is the functional derivative, with respect to arbitrary norm conserving variations of the density, of a yet unknown 'electron-interaction' energy functional in which *all* the electron correlations are incorporated. For an external potential that is local, the electrons then move in the *same* local effective potential. The 'electron-interaction' potential of Kohn–Sham theory is, of course, unknown. However, the *a priori* knowledge and understanding of the structure of this potential is further obscured by its definition as a functional derivative. Thus, what is required is a fundamental *physical* interpretation of this local potential in which all the many-body effects are incorporated. (It is noted, however, that the asymptotic structure of the 'electron-interaction' potential and of its exchange and correlations components in the classically forbidden region of finite [5–9] nonuniform electron density systems such as atoms, molecules, and metallic clusters, and extended [10] systems such as metallic surfaces, is at present known analytically).

In this article we describe the *rigorous quantum-mechanical* interpretation of the 'electron-interaction' potential of Kohn–Sham theory based on the original ideas of Harbola and Sahni [9], and of their subsequent extension by

Holas and March [11]. This physical description further distinguishes between the Pauli and Coulomb correlation components and the correlation-kinetic-energy component of the potential, and thereby provides insights into its structure and of its components. In addition the interpretation helps distinguish between the Pauli, Coulomb and correlation-kinetic-energy contributions to the electron-interaction energy functional. A consistent physical description for both the energy functional and its functional derivative is thereby achieved.

For the quantum-mechanical interpretation of density-functional theory, we begin with the definitions of requisite properties within Schrödinger and Kohn–Sham theories. We then provide a description and proof of the physical interpretation of the local 'electron-interaction' potential of Kohn–Sham theory. Following this we show how the interpretation leads to definitions of the Pauli, Coulomb, and correlation-kinetic-energy components of the electron-interaction energy and local potential in terms of respective quantum-mechanical source charge distributions and their fields. This in turn leads to a discussion of the structure of these components of the potential in the classically forbidden region, and thereby to the asymptotic structure of the electron-interaction potential for finite and extended systems. Recent results of application to the ground and excited states of atoms are presented. Lastly, a few important consequences of the physical interpretation are discussed. We comment on the Hohenberg–Kohn theorem in terms of the inverse map whereby wavefunctions lead to external potentials for both non-degenerate and degenerate ground states. We then explain the electronic theory of Slater [12], the precursor to modern density-functional theory, in terms of the new understandings achieved. Finally, we explain by example, how the physical interpretation bears [13] on approximate Kohn–Sham theory. When the 'electron-interaction' energy functional is approximated, and the potential obtained as its functional derivative, it is assumed that the electrons are correlated as in the definition of the approximate functional. However, when the requisite quantum-mechanical source charge for the approximation is determined, and the expressions for the energy and potential rederived via the physical interpretation, correlations beyond those assumed in the construction of the approximate energy functional emerge. The knowledge of these additional correlations then leads to a more meaningful evaluation of the results of the approximation. We conclude with a discussion of how the physical interpretation leads to the *a priori* understanding of the general structure of the 'electron-interaction' potential as well as what yet remains to be understood. We subsequently end with remarks for future work.

## 2 Schrödinger and Kohn–Sham Theories

For the quantum-mechanical description of Kohn–Sham density-functional theory, we define in this section properties within the context of Schrödinger theory relevant to the interpretation. We also give a brief description of

Kohn–Sham theory in order to define the local potential representing electron correlations as well as other properties derived within its context.

## 2.1 Definitions within Schrödinger Theory

The Hamiltonian $\hat{H}$ of a system of N electrons in an external potential represented by an operator $\hat{V}$ of the local single-particle form is

$$\hat{H} = \hat{T} + \hat{V} + \hat{U} \tag{1}$$

where the kinetic energy operator

$$\hat{T} = \sum_i -\frac{1}{2}\nabla_i^2 \tag{2}$$

the external potential operator

$$\hat{V} = \sum_i v(\mathbf{r}_i) \tag{3}$$

and the electron-interaction operator

$$\hat{U} = \frac{1}{2}\sum_{i,j}{}' \frac{1}{|\mathbf{r}_i - \mathbf{r}_j|} . \tag{4}$$

The Schrödinger equation is then

$$\hat{H}\Psi(\mathbf{x}_1,\dots\mathbf{x}_N) = E\Psi(\mathbf{x}_1,\dots\mathbf{x}_N) \tag{5}$$

where $\Psi$ and E are the normalized system wavefunction and energy, respectively. The energy is the expectation $E = \langle\Psi|\hat{H}|\Psi\rangle$. (Here $\mathbf{x} = \mathbf{r}\sigma$, where $\mathbf{r}$ is the spatial and $\sigma$ the spin coordinate of the electron. The integral $\int d\mathbf{x} \equiv \Sigma_\sigma \int d\mathbf{r}$).

The first property of interest is the *spinless single-particle density matrix* $\gamma(\mathbf{r}, \mathbf{r}')$ defined as

$$\gamma(\mathbf{r}, \mathbf{r}') = N\sum_\sigma \int \Psi^*(\mathbf{r}\sigma, \mathbf{x}_2,\dots\mathbf{x}_N)\Psi(\mathbf{r}'\sigma, \mathbf{x}_2, \dots\mathbf{x}_N)\,d\mathbf{x}_2,\dots d\mathbf{x}_N. \tag{6}$$

This density matrix can also be written [14] as the expectation value of the Hermitian operator $\hat{X}$:

$$\gamma(\mathbf{r}, \mathbf{r}') = \langle\Psi|\hat{X}|\Psi\rangle \tag{7}$$

where

$$\hat{X} = \hat{A} + i\hat{B} \tag{8}$$

$$\hat{A} = \frac{1}{2}\sum_j [\delta(\mathbf{r}_j - \mathbf{r})T_j(\mathbf{a}) + \delta(\mathbf{r}_j - \mathbf{r}')T_j(-\mathbf{a})] \tag{9}$$

$$\hat{B} = \frac{i}{2}\sum_j [\delta(\mathbf{r}_j - \mathbf{r})T_j(\mathbf{a}) - \delta(\mathbf{r}_j - \mathbf{r}')T_j(-\mathbf{a})] . \tag{10}$$

$T_j(\mathbf{a})$ is a translation operator such that $T_j(\mathbf{a})\Psi(\ldots \mathbf{r}_j \ldots) = \Psi(\ldots \mathbf{r}_j + \mathbf{a} \ldots)$, and $\mathbf{a} = \mathbf{r}' - \mathbf{r}$. The single-particle density matrix constructed from the wavefunction $\Psi$ is not idempotent and satisfies the condition

$$\int d\mathbf{x}'' \gamma(\mathbf{x}, \mathbf{x}'') \gamma(\mathbf{x}'', \mathbf{x}') < \gamma(\mathbf{x}, \mathbf{x}') . \tag{11}$$

The diagonal matrix element of the density matrix is the density $\rho(\mathbf{r})$. Equivalently, it is the expectation value of the density operator

$$\hat{\rho}(\mathbf{r}) = \sum_i \delta(\mathbf{r}_i - \mathbf{r}) \tag{12}$$

so that

$$\rho(\mathbf{r}) = \gamma(\mathbf{r}, \mathbf{r}) = \langle \Psi | \hat{\rho} | \Psi \rangle . \tag{13}$$

The property associated [9] with the purely electron-interaction component of the Kohn–Sham theory many-body potential as well as the electron-interaction energy is the *pair-correlation density* $g(\mathbf{r}, \mathbf{r}')$. It is defined in terms of the pair-correlation operator

$$\hat{P}(\mathbf{r}, \mathbf{r}') = \sum_{i,j}{}' \delta(\mathbf{r}_i - \mathbf{r}) \delta(\mathbf{r}_j - \mathbf{r}') \tag{14}$$

as

$$g(\mathbf{r}, \mathbf{r}') = \langle \Psi | \hat{P}(\mathbf{r}, \mathbf{r}') | \Psi \rangle / \rho(\mathbf{r}) . \tag{15}$$

Note that in the definition of the pair-correlation density there is no self-interaction. In physical terms, the pair-correlation density is the *density* at $\mathbf{r}'$ for an electron at $\mathbf{r}$. Its total charge for arbitrary electron position is thus

$$\int g(\mathbf{r}, \mathbf{r}') d\mathbf{r}' = N - 1 . \tag{16}$$

The pair-correlation density is a property that arises due to the Pauli and Coulomb correlations between electrons. Thus it can also be interpreted as the density $\rho(\mathbf{r}')$ at $\mathbf{r}'$ plus the reduction in this density at $\mathbf{r}'$ due to the electron correlations. The reduction in density about an electron which occurs as a result of the Pauli exclusion principle and Coulomb repulsion is the *quantum-mechanical Fermi-Coulomb hole* charge distribution $\rho_{xc}(\mathbf{r}, \mathbf{r}')$. Thus we may write the pair-correlation density as

$$g(\mathbf{r}, \mathbf{r}') = \rho(\mathbf{r}') + \rho_{xc}(\mathbf{r}, \mathbf{r}') \tag{17}$$

and consequently the total charge of the Fermi–Coulomb hole for arbitrary electron position is

$$\int \rho_{xc}(\mathbf{r}, \mathbf{r}') d\mathbf{r}' = -1 . \tag{18}$$

Note that the self-interaction contribution to the Fermi–Coulomb hole charge is cancelled by the density, so that the pair-correlation density as defined by Eq. (17) is self-interaction free.

The *electron-interaction energy* $E_{ee}$, which is the expectation value of the operator $\hat{U}$, can be afforded a physical interpretation in terms of the pair-correlation density as the energy of interaction between it and the electronic density:

$$E_{ee} = \langle \Psi | \hat{U} | \Psi \rangle = \frac{1}{2} \int \int \frac{\rho(\mathbf{r}) g(\mathbf{r}, \mathbf{r}')}{|\mathbf{r} - \mathbf{r}'|} d\mathbf{r}\, d\mathbf{r}' . \tag{19}$$

Using the form of $g(\mathbf{r}, \mathbf{r}')$ as given by Eq. (17), the electron-interaction energy can be split further as

$$E_{ee} = E_H + E_{xc} \tag{20}$$

where $E_H$ is the Coulomb self-energy

$$E_H = \frac{1}{2} \int \int \frac{\rho(\mathbf{r}) \rho(\mathbf{r}')}{|\mathbf{r} - \mathbf{r}'|} d\mathbf{r}\, d\mathbf{r}' \tag{21}$$

and $E_{xc}$ is the *quantum-mechanical exchange-correlation energy*

$$E_{xc} = \frac{1}{2} \int \int \frac{\rho(\mathbf{r}) \rho_{xc}(\mathbf{r}, \mathbf{r}')}{|\mathbf{r} - \mathbf{r}'|} d\mathbf{r}\, d\mathbf{r}' \tag{22}$$

which is the energy of interaction between the density and the Fermi-Coulomb hole charge distribution.

The property associated[11] with the correlation-kinetic-energy component of the Kohn-Sham potential is the *kinetic-energy-density tensor* $t_{\alpha\beta}(\mathbf{r})$. This is a real, symmetric tensor defined in terms of the single-particle density matrix $\gamma(\mathbf{r}, \mathbf{r}')$ as

$$t_{\alpha\beta}(\mathbf{r}; [\gamma]) = \frac{1}{4} \left[ \frac{\partial^2}{\partial r'_\alpha \partial r''_\beta} + \frac{\partial^2}{\partial r'_\beta \partial r''_\alpha} \right] \gamma(\mathbf{r}', \mathbf{r}'')|_{\mathbf{r}' = \mathbf{r}'' = \mathbf{r}} . \tag{23}$$

The trace of the kinetic-energy-density tensor is the scalar *kinetic energy density* $t(\mathbf{r})$:

$$t(\mathbf{r}) = \sum_\alpha t_{\alpha\alpha}(\mathbf{r}) \geq 0 . \tag{24}$$

The *kinetic energy* T, which is the expectation value of the operator $\hat{T}$, is then

$$T = \langle \Psi | \hat{T} | \Psi \rangle = \int d\mathbf{r}\, t(\mathbf{r})$$

$$= \frac{1}{2} \int d\mathbf{r}\, [\nabla_\mathbf{r} \cdot \nabla_{\mathbf{r}'} \gamma(\mathbf{r}, \mathbf{r}')]_{\mathbf{r}' = \mathbf{r}} . \tag{25}$$

Finally, the total energy E can thus be written as

$$E = T + \int v(\mathbf{r})\rho(\mathbf{r})\,d\mathbf{r} + E_{ee}$$

$$= T + \int v(\mathbf{r})\rho(\mathbf{r})d\mathbf{r} + E_H + E_{xc} \qquad (26)$$

with T, $E_H$, and $E_{xc}$ as defined above.

## 2.2 Definitions Within Kohn–Sham Theory

The basic idea underlying Kohn–Sham theory [4] is the construction of a model system of noninteracting quasi-particles for which the density is the same as that of the interacting system. As such the ground-state energy functional $E[\rho]$ is partitioned as

$$E[\rho] = T_s[\rho] + \int v(\mathbf{r})\rho(\mathbf{r})\,d\mathbf{r} + E_{ee}^{KS}[\rho] \qquad (27)$$

where $T_s[\rho]$ is the corresponding kinetic energy of the noninteracting system. This equation defines the Kohn–Sham theory electron-interaction energy functional $E_{ee}^{KS}[\rho]$ which can then be further partitioned as

$$E_{ee}^{KS}[\rho] = E_H[\rho] + E_{xc}^{KS}[\rho] \qquad (28)$$

where $E_H[\rho]$ is the Coulomb self-energy defined previously. Comparison with Eq. (26) for the energy expression in Schrödinger theory then defines the Kohn–Sham theory exchange-correlation energy functional $E_{xc}^{KS}[\rho]$ as the sum of the quantum-mechanical exchange-correlation energy $E_{xc}$ and the correlation-kinetic-energy $T_c[\rho]$:

$$E_{xc}^{KS}[\rho] = E_{xc}[\rho] + T_c[\rho] \qquad (29)$$

where in turn

$$T_c[\rho] = T[\rho] - T_s[\rho] . \qquad (30)$$

The application of the variational principle to the ground-state energy functional of Eq. (27) for arbitrary norm conserving variations of the density leads to the Kohn–Sham equation

$$\left[ -\frac{1}{2}\nabla^2 + v(\mathbf{r}) + v_{ee}^{KS}(\mathbf{r}) \right]\phi_i(\mathbf{x}) = \varepsilon_i\phi_i(\mathbf{x}); \quad i = 1,\dots N \qquad (31)$$

where $v_{ee}^{KS}(\mathbf{r})$ is the local potential in which all the electron correlations are incorporated. As a result of the variational principle, this potential is derived to be the functional derivative of $E_{ee}^{KS}[\rho]$:

$$v_{ee}^{KS}(\mathbf{r}) = \frac{\delta E_{ee}^{KS}[\rho]}{\delta\rho(\mathbf{r})} . \qquad (32)$$

With the partition of $E_{ee}^{KS}[\rho]$ according to Eq. (28), the potential can be written as the sum

$$v_{ee}^{KS}(\mathbf{r}) = v_H(\mathbf{r}) + v_{xc}^{KS}(\mathbf{r}) \tag{33}$$

which defines the density-functional theory Hartree potential $v_H(\mathbf{r})$ as the functional derivative

$$v_H(\mathbf{r}) = \frac{\delta E_H[\rho]}{\delta\rho(\mathbf{r})} = \int \frac{\rho(\mathbf{r}')}{|\mathbf{r} - \mathbf{r}'|} d\mathbf{r}' \tag{34}$$

and the Kohn–Sham theory 'exchange-correlation' potential $v_{xc}^{KS}(\mathbf{r})$ as the functional derivative

$$v_{xc}^{KS}(\mathbf{r}) = \frac{\delta E_{xc}^{KS}[\rho]}{\delta\rho(\mathbf{r})}. \tag{35}$$

The ground-state 'wavefunction' corresponding to this noninteracting system is then a single Slater determinant $\Phi_s\{\phi_i(\mathbf{x})\}$ of the lowest occupied orbitals $\phi_i(\mathbf{x})$ of the Kohn–Sham differential equation. The Dirac [15] single-particle density matrix $\gamma_s(\mathbf{r}, \mathbf{r}')$ that results from this Slater determinant is

$$\gamma_s(\mathbf{r}, \mathbf{r}') = \sum_i \sum_\sigma \phi_i^*(\mathbf{r}\sigma)\phi_i(\mathbf{r}'\sigma) \tag{36}$$

and it is idempotent:

$$\int \gamma_s(\mathbf{x}, \mathbf{x}'')\gamma_s(\mathbf{x}'', \mathbf{x}') d\mathbf{x}'' = \gamma_s(\mathbf{x}, \mathbf{x}'). \tag{37}$$

The exact ground-state density $\rho(\mathbf{r})$ and the noninteracting kinetic energy $T_s[\rho]$ are also obtained from this Slater determinant as

$$\rho(\mathbf{r}) = \sum_i \sum_\sigma |\phi_i(\mathbf{r}\sigma)|^2 \tag{38}$$

and

$$T_s[\rho] = \sum_i \int \phi_i^*(\mathbf{x})\left[ -\frac{1}{2}\nabla^2 \right]\phi_i(\mathbf{x}) d\mathbf{r} \tag{39}$$

respectively. The ground-state energy is then determined by the energy functional of Eq. (27). Finally, in addition to generating the orbitals from which the exact ground-state density and energy of the interacting system are determined, the highest occupied eigenvalue of the Kohn–Sham differential equation of Eq. (31) has the physical interpretation [16] of being the removal energy. Thus, in principle, its solution can lead to the determination of properties such as the ionization potential, electron affinity and work function.

## 3 Physical Interpretation of Electron-Interaction Potential of Kohn–ShamTheory

Since the electron-interaction energy functional $E_{ee}^{KS}[\rho]$ of Kohn–Sham theory is representative of Pauli and Coulomb correlations as well as the correlation contribution to the kinetic energy, so is the corresponding local potential $v_{ee}^{KS}(\mathbf{r})$ obtained from it through functional differentiation. In the physical interpretation of the potential $v_{ee}^{KS}(\mathbf{r})$, however, it is possible to distinguish between the *purely* quantum-mechanical (Pauli and Coulomb) electron-correlation component $W_{ee}(\mathbf{r})$, and the correlation-kinetic-energy component $W_{t_c}(\mathbf{r})$. We begin this section with a description of the physical interpretation of $v_{ee}^{KS}(\mathbf{r})$, and then discuss its components $W_{ee}(\mathbf{r})$ and $W_{t_c}(\mathbf{r})$ more fully.

The electron-interaction potential $v_{ee}^{KS}(\mathbf{r})$ of Kohn–Sham theory is the work done to bring an electron from infinity to its position at $\mathbf{r}$ against a field $\mathscr{F}(\mathbf{r})$:

$$v_{ee}^{KS}(\mathbf{r}) = \frac{\delta E_{ee}^{KS}[\rho]}{\delta\rho(\mathbf{r})} = -\int_{\infty}^{\mathbf{r}} \mathscr{F}(\mathbf{r}')\cdot d\mathbf{l}' . \tag{40}$$

The field $\mathscr{F}(\mathbf{r})$ is the sum of two fields:

$$\mathscr{F}(\mathbf{r}) = \mathscr{E}_{ee}(\mathbf{r}) + Z_{t_c}(\mathbf{r}) . \tag{41}$$

The field $\mathscr{E}_{ee}(\mathbf{r})$ is strictly representative of Pauli and Coulomb correlations since its quantum-mechanical source charge distribution is the pair-correlation density $g(\mathbf{r},\mathbf{r}')$. On the other hand, the field $Z_{t_c}(\mathbf{r})$ arises from the kinetic-energy-density tensor $t_{\alpha\beta}(\mathbf{r})$. It is the difference of the fields derived from the tensor for the interacting and Kohn–Sham noninteracting systems, and is thereby representative of the correlation-kinetic-energy.

Thus the potential $v_{ee}^{KS}(\mathbf{r})$ may be written as

$$v_{ee}^{KS}(\mathbf{r}) = W_{ee}(\mathbf{r}) + W_{t_c}(\mathbf{r}) \tag{42}$$

where

$$W_{ee}(\mathbf{r}) = -\int_{\infty}^{\mathbf{r}} \mathscr{E}_{ee}(\mathbf{r}')\cdot d\mathbf{l}' \tag{43}$$

and

$$W_{t_c}(\mathbf{r}) = -\int_{\infty}^{\mathbf{r}} Z_{t_c}(\mathbf{r}')\cdot d\mathbf{l}' . \tag{44}$$

The interpretation of the functional derivative $v_{ee}^{KS}(\mathbf{r})$ as the work done is due to the fact that it can be written as

$$\nabla v_{ee}^{KS}(\mathbf{r}) = -\mathscr{F}(\mathbf{r}) \tag{45}$$

so that the sum of the work $W_{ee}(\mathbf{r})$ and $W_{t_c}(\mathbf{r})$ is path-independent. The path-independence of the work is, of course, rigorously valid provided the field

$\mathscr{F}(\mathbf{r})$ is smooth, i.e. it is continuous, differentiable, and has continuous first derivatives. Equation (45) also implies that the curl of the field $\mathscr{F}(\mathbf{r})$ vanishes:

$$\nabla \times \mathscr{F}(\mathbf{r}) = 0 . \tag{46}$$

For systems of a certain symmetry such as closed shell atoms, jellium metal clusters, jellium metal surfaces, open-shell atoms in the central-field approximation, etc., the work $W_{ee}(\mathbf{r})$ and $W_{t_c}(\mathbf{r})$ are separately path-independent since $\nabla \times \mathscr{E}_{ee}(\mathbf{r}) = \nabla \times Z_{t_c}(\mathbf{r}) = 0$.

## 3.1 The Quantum-Mechanical Electron-Interaction Component $W_{ee}(\mathbf{r})$

The physical interpretation of the electron-interaction component $W_{ee}(\mathbf{r})$ was originally proposed by Harbola and Sahni [9], and derived by them via Coulomb's law. It is based on the observation that the pair-correlation density $g(\mathbf{r}, \mathbf{r}')$ is not a static but rather a *dynamic* charge distribution whose structure changes as a function of electron position. The dynamic nature of this charge then must be accounted for in the description of the potential. Thus, in order to obtain the local potential in which the electron moves, the force field $\mathscr{E}_{ee}(\mathbf{r})$ due to this charge distribution must first be determined. According to Coulomb's law this field is

$$\mathscr{E}_{ee}(\mathbf{r}) = \int \frac{g(\mathbf{r}, \mathbf{r}')(\mathbf{r} - \mathbf{r}')}{|\mathbf{r} - \mathbf{r}'|^3} \, d\mathbf{r}' . \tag{47}$$

The component $W_{ee}(\mathbf{r})$ is then the work done to bring an electron from infinity to its position at $\mathbf{r}$ in this force field as defined by Eq. (43).

The component $W_{ee}(\mathbf{r})$ can be further simplified by employing the expression for $g(\mathbf{r}, \mathbf{r}')$ (see Eq. (17)) in terms of the density $\rho(\mathbf{r}')$ and the Fermi-Coulomb hole charge density $\rho_{xc}(\mathbf{r}, \mathbf{r}')$. The field $\mathscr{E}_{ee}(\mathbf{r})$ is then the sum of the Hartree $\mathscr{E}_H(\mathbf{r})$ and exchange-correlation $\mathscr{E}_{xc}(\mathbf{r})$ fields:

$$\mathscr{E}_{ee}(\mathbf{r}) = \mathscr{E}_H(\mathbf{r}) + \mathscr{E}_{xc}(\mathbf{r}) \tag{48}$$

where

$$\mathscr{E}_H(\mathbf{r}) = \int \frac{\rho(\mathbf{r}')(\mathbf{r} - \mathbf{r}')}{|\mathbf{r} - \mathbf{r}'|^3} \, d\mathbf{r}' \quad \text{and} \quad \mathscr{E}_{xc}(\mathbf{r}) = \int \frac{\rho_{xc}(\mathbf{r}, \mathbf{r}')(\mathbf{r} - \mathbf{r}')}{|\mathbf{r} - \mathbf{r}'|^3} \, d\mathbf{r}' . \tag{49}$$

The component $W_{ee}(\mathbf{r})$ is in turn the sum of the work done $W_H(\mathbf{r})$ and $W_{xc}(\mathbf{r})$ to move an electron in the Hartree and exchange-correlation fields respectively:

$$W_{ee}(\mathbf{r}) = W_H(\mathbf{r}) + W_{xc}(\mathbf{r}) \tag{50}$$

where

$$W_H(\mathbf{r}) = - \int_\infty^\mathbf{r} \mathscr{E}_H(\mathbf{r}') \cdot d\mathbf{l}' \quad \text{and} \quad W_{xc}(\mathbf{r}) = - \int_\infty^\mathbf{r} \mathscr{E}_{xc}(\mathbf{r}') \cdot d\mathbf{l}' . \tag{51}$$

Now the electronic density $\rho(\mathbf{r})$ is a *static* charge distribution whose structure does not change as a function of electron position. Thus the Hartree field can be written as $\mathscr{E}_H(\mathbf{r}) = -\nabla W_H(\mathbf{r})$, where

$$W_H(\mathbf{r}) = \int \frac{\rho(\mathbf{r}')}{|\mathbf{r} - \mathbf{r}'|} d\mathbf{r}' . \tag{52}$$

The work $W_H(\mathbf{r})$ is path-independent and $\nabla \times \mathscr{E}_H(\mathbf{r}) = 0$. Furthermore, the scalar potential $W_H(\mathbf{r})$ is recognized to be the density-functional theory Hartree potential $v_H(\mathbf{r})$ of Eq. (34). *Thus the functional derivative of the Coulomb self-energy functional $E_H[\rho]$ has the physical interpretation of being the work done in the field of the electronic density. The component $W_{ee}(\mathbf{r})$ is then the sum of the Hartree potential and the work done to move an electron in the field of the quantum-mechanical Fermi-Coulomb hole charge distribution:*

$$W_{ee}(\mathbf{r}) = W_H(\mathbf{r}) + W_{xc}(\mathbf{r}) . \tag{53}$$

This work $W_{xc}(\mathbf{r})$ is path-independent for the symmetrical density systems noted previously since the $\nabla \times \mathscr{E}_{xc}(\mathbf{r}) = 0$ for these cases. It is important to note, however, that the corresponding Fermi–Coulomb hole charge distribution $\rho_{xc}(\mathbf{r}, \mathbf{r}')$ which gives rise to the field $\mathscr{E}_{xc}(\mathbf{r})$ need not possess the same symmetry for arbitrary electron position. For example, in either closed shell atoms or open-shell atoms in the central-field approximation for which the density is spherically symmetric, the Fermi–Coulomb hole is not, the only exception being when the electron is at the nucleus.

## 3.2 The Correlation-Kinetic-Energy Component $W_{t_c}(\mathbf{r})$

The correlation-kinetic-energy component $W_{t_c}(\mathbf{r})$ is the work done to move an electron in the field $Z_{t_c}(\mathbf{r})$ as expressed by Eq. (44). The field $Z_{t_c}(\mathbf{r})$ is given in terms of a field $\mathbf{z}(\mathbf{r}; [\gamma])$ whose component $z_\alpha(\mathbf{r})$ is derived from the kinetic-energy-density tensor $t_{\alpha\beta}(\mathbf{r}; [\gamma])$ as

$$z_\alpha(\mathbf{r}; [\gamma]) = 2 \sum_{\beta=1}^{3} \frac{\partial}{\partial r_\beta} t_{\alpha\beta}(\mathbf{r}; [\gamma]) . \tag{54}$$

The field $\mathbf{z}(\mathbf{r}; [\gamma])$ thus defined is for the interacting system since the tensor involves the density matrix $\gamma(\mathbf{r}, \mathbf{r}')$ of Eq. (6). With the field $\mathbf{z}(\mathbf{r}; [\gamma_s])$ derived similarly from the tensor $t_{\alpha\beta}(\mathbf{r}; [\gamma_s])$ written in terms of the idempotent Dirac density matrix $\gamma_s(\mathbf{r}, \mathbf{r}')$ of Kohn–Sham theory, the field $Z_{t_c}(\mathbf{r})$ is then defined as

$$Z_{t_c}(\mathbf{r}) = \frac{1}{\rho(\mathbf{r})} [\mathbf{z}(\mathbf{r}; [\gamma_s]) - \mathbf{z}(\mathbf{r}; [\gamma])] . \tag{55}$$

Note that the determination of this field thus requires knowledge of the Kohn–Sham orbitals.

## 3.3 Proof via the Virial Theorem

The electron-interaction component $W_{ee}(r)$ was originally derived, as noted previously, by Harbola and Sahni [9] via Coulomb's law. Since this component does not contain any correlation-kinetic-energy contributions, it does not [9, 17, 18] satisfy the Kohn–Sham theory sum rule relating the corresponding electron-correlation energy $E_{ee}^{KS}[\rho]$ to its functional derivative (potential) $v_{ee}^{KS}(r)$. The sum rule, which is derived [19, 20] from the virial theorem, and in which the correlation-kinetic-energy $T_c[\rho]$ contribution is made explicit is

$$E_{ee}^{KS}[\rho] + \int dr \rho(r) r \cdot \nabla v_{ee}^{KS}(r) = - T_c \leq 0 . \tag{56}$$

Consequently, Harbola and Sahni [9, 18] proposed that a term which accounts for the correlation-kinetic-energy contribution be added to $W_{ee}(r)$ in order to obtain the Kohn–Sham potential $v_{ee}^{KS}(r)$. This term is the work $W_{t_c}(r)$. Both the components $W_{ee}(r)$ and $W_{t_c}(r)$ can, however, be derived from the virial theorem and we give here the proof according to Holas and March [11].

The integral form of the quantum-mechanical virial theorem which is

$$2T + E_{ee} = \int dr \rho(r) r \cdot \nabla v(r) \tag{57}$$

can be written in differential form [11] as

$$\nabla v(r) = - F(r) \tag{58}$$

where

$$F(r) = - \mathscr{E}_{ee}(r) + \frac{1}{\rho(r)} [ - \tfrac{1}{4} \nabla \nabla^2 \rho(r) + z(r; [\gamma])] . \tag{59}$$

Note that the field $F(r)$ depends upon the density $\rho(r)$, as well as the single-particle density matrix $\gamma(r, r')$ and the pair-correlation density $g(r, r')$ through the fields $z(r)$ and $\mathscr{E}_{ee}(r)$, respectively. The corresponding differential form of the virial theorem for the noninteracting Kohn–Sham system is

$$\nabla v_s(r) = - F_s(r) \tag{60}$$

where (see Eq. (31))

$$v_s(r) = v(r) + v_{ee}^{KS}(r) \tag{61}$$

and

$$F_s(r) = \frac{1}{\rho(r)} [ - \tfrac{1}{4} \nabla \nabla^2 \rho(r) + z(r; [\gamma_s])] . \tag{62}$$

The field $F_s(r)$ depends only on the density $\rho(r)$ and the idempotent density matrix $\gamma_s(r, r')$ through the field $z(r; [\gamma_s])$. The field $\mathscr{E}_{ee}(r)$ does not appear in the expression for $F_s(r)$ because there is no electron-interaction operator in the

Kohn–Sham differential equation. On subtracting Eq. (58) from Eq. (60) one obtains

$$\nabla v_{ee}^{KS}(\mathbf{r}) = - \left\{ \mathscr{E}_{ee}(\mathbf{r}) + \frac{1}{\rho(\mathbf{r})} \left[ \mathbf{z}(\mathbf{r}; [\gamma_s]) - \mathbf{z}(\mathbf{r}; [\gamma]) \right] \right\}$$

$$= - \mathscr{F}(\mathbf{r}) \tag{63}$$

which in turn leads to the interpretation of $v_{ee}^{KS}(\mathbf{r})$ as the work done to move an electron in the field $\mathscr{F}(\mathbf{r})$.

# 4 Further Definitions Within Kohn–Sham Theory

According to the physical interpretation, the Kohn–Sham theory electron-interaction potential $v_{ee}^{KS}[\rho]$ is the sum

$$v_{ee}^{KS}[\rho] = W_H(\mathbf{r}) + W_{xc}(\mathbf{r}) + W_{t_c}(\mathbf{r}) \tag{64}$$

where $W_H(\mathbf{r})$ is the work done in the Hartree field $\mathscr{E}_H(\mathbf{r})$ arising from the density $\rho(\mathbf{r})$, $W_{xc}(\mathbf{r})$ is the work done in the exchange-correlation field $\mathscr{E}_{xc}(\mathbf{r})$ due to the quantum-mechanical Fermi-Coulomb hole charge $\rho_{xc}(\mathbf{r}, \mathbf{r}')$, and $W_{t_c}(\mathbf{r})$ is the work done in the field $\mathbf{Z}_{t_c}(\mathbf{r})$ derived from the kinetic-energy-density tensor $t_{\alpha\beta}(\mathbf{r})$. The corresponding Kohn–Sham theory electron-interaction energy $E_{ee}^{KS}[\rho]$ is the sum

$$E_{ee}^{KS}[\rho] = E_H[\rho] + E_{xc}[\rho] + T_c[\rho] \tag{65}$$

where $E_H[\rho]$ is the Coulomb self-energy, $E_{xc}[\rho]$ is the quantum-mechanical exchange-correlation energy, and $T_c[\rho]$ is the correlation-kinetic-energy.

We next further split the quantum-mechanical quantities $\rho_{xc}(\mathbf{r})$, $E_{xc}[\rho]$ and $W_{xc}(\mathbf{r})$ into their Kohn–Sham theory exchange and resulting correlation components. In this manner it is then possible to write each component of $E_{ee}^{KS}[\rho]$ in terms of the corresponding field which gives rise to it. It also allows for an understanding of the structure of each component of the electron-interaction potential $v_{ee}^{KS}(\mathbf{r})$. (We note that there are various other [21] definitions of the exchange and correlation components of the Kohn–Sham theory exchange-correlation energy and potential employed in the literature.)

## 4.1 Kohn–Sham Theory Fermi and Coulomb Holes, and Exchange and Correlation Energies

The pair-correlation density $g_s(\mathbf{r}, \mathbf{r}')$ derived from the Kohn–Sham theory Slater determinant $\Phi_s\{\phi_i\}$ is

$$g_s(\mathbf{r}, \mathbf{r}') = \langle \Phi_s\{\phi_i\} | \hat{P}(\mathbf{r}, \mathbf{r}') | \Phi_s\{\phi_i\} \rangle / \rho(\mathbf{r}) \tag{66}$$

and defines the Kohn–Sham theory Fermi hole $\rho_x^{KS}(\mathbf{r}, \mathbf{r}')$ since $g_s(\mathbf{r}, \mathbf{r}')$ can also be written as

$$g_s(\mathbf{r}, \mathbf{r}') = \rho(\mathbf{r}') + \rho_x^{KS}(\mathbf{r}, \mathbf{r}') \ . \tag{67}$$

The Fermi hole $\rho_x^{KS}(\mathbf{r}, \mathbf{r}')$ in turn is defined in terms of the idempotent Dirac density matrix $\gamma_s(\mathbf{r}, \mathbf{r}')$ as

$$\rho_x^{KS}(\mathbf{r}, \mathbf{r}') = - |\gamma_s(\mathbf{r}, \mathbf{r}')|^2 / 2\rho(\mathbf{r}) \tag{68}$$

and satisfies the constraints of charge neutrality, negativity, and value at electron position:

$$\int \rho_x^{KS}(\mathbf{r}, \mathbf{r}') \, d\mathbf{r}' = -1 \tag{69}$$

$$\rho_x^{KS}(\mathbf{r}, \mathbf{r}') \leqslant 0 \tag{70}$$

$$\rho_x^{KS}(\mathbf{r}, \mathbf{r}) = \rho(\mathbf{r})/2 \ . \tag{71}$$

The corresponding Kohn–Sham exchange energy $E_x^{KS}[\rho]$ is the energy of interaction between the density $\rho(\mathbf{r})$ and the Fermi hole $\rho_x^{KS}(\mathbf{r}, \mathbf{r}')$:

$$E_x^{KS}[\rho] = \frac{1}{2} \int \int \frac{\rho(\mathbf{r})\rho_x^{KS}(\mathbf{r}, \mathbf{r}')}{|\mathbf{r} - \mathbf{r}'|} \, d\mathbf{r} \, d\mathbf{r}' \ . \tag{72}$$

We define the Kohn–Sham theory Coulomb hole $\rho_c^{KS}(\mathbf{r}, \mathbf{r}')$ as the difference

$$\rho_c^{KS}(\mathbf{r}, \mathbf{r}') = \rho_{xc}(\mathbf{r}, \mathbf{r}') - \rho_x^{KS}(\mathbf{r}, \mathbf{r}') \tag{73}$$

so that it satisfies the constraint

$$\int \rho_c^{KS}(\mathbf{r}, \mathbf{r}') \, d\mathbf{r}' = 0 \ . \tag{74}$$

This definition ensures that together with the Kohn–Sham theory Fermi $\rho_x^{KS}(\mathbf{r}, \mathbf{r}')$, and the quantum-mechanical Fermi–Coulomb $\rho_{xc}(\mathbf{r}, \mathbf{r}')$ holes, the Coulomb hole $\rho_c^{KS}(\mathbf{r}, \mathbf{r}')$ too corresponds to the system density $\rho(\mathbf{r})$. The correlation energy $E_c^{KS}[\rho]$ is then

$$E_c^{KS}[\rho] = U_c[\rho] + T_c[\rho] \tag{75}$$

where

$$U_c[\rho] = \frac{1}{2} \int \int \frac{\rho(\mathbf{r})\rho_c^{KS}(\mathbf{r}, \mathbf{r}')}{|\mathbf{r} - \mathbf{r}'|} \, d\mathbf{r} \, d\mathbf{r}' \tag{76}$$

is the energy of interaction between the densities $\rho(\mathbf{r})$ and $\rho_c^{KS}(\mathbf{r}, \mathbf{r}')$. Thus the Kohn–Sham exchange-correlation energy $E_{xc}^{KS}[\rho]$ is the sum

$$E_{xc}^{KS}[\rho] = E_x^{KS}[\rho] + E_c^{KS}[\rho] \tag{77}$$

with $E_x^{KS}[\rho]$ and $E_c^{KS}[\rho]$ as defined above.

## 4.2 Kohn–Sham Theory Exchange and Correlation Fields and Potentials

In standard Kohn–Sham theory, the exchange-correlation potential $v_{xc}^{KS}(\mathbf{r})$ is the sum of its exchange $v_x^{KS}(\mathbf{r})$ and correlation $v_c^{KS}(\mathbf{r})$ components which are in turn defined as the functional derivatives

$$v_x^{KS}(\mathbf{r}) = \delta E_x^{KS}[\rho]/\delta\rho(\mathbf{r}) \tag{78}$$

and

$$v_c^{KS}(\mathbf{r}) = \delta E_c^{KS}[\rho]/\delta\rho(\mathbf{r}) \tag{79}$$

respectively. However, with the Fermi $\rho_x^{KS}(\mathbf{r}, \mathbf{r}')$ and Coulomb $\rho_c^{KS}(\mathbf{r}, \mathbf{r}')$ holes as defined in the previous section, the potential $v_{xc}^{KS}(\mathbf{r})$ can be written as

$$v_{xc}^{KS}(\mathbf{r}) = \frac{\delta E_{xc}^{KS}[\rho]}{\delta\rho(\mathbf{r})} = W_x^{KS}(\mathbf{r}) + W_c^{KS}(\mathbf{r}) + W_{t_c}(\mathbf{r}) \tag{80}$$

where $W_{t_c}(\mathbf{r})$ is the work done in the field $\mathbf{Z}_{t_c}(\mathbf{r})$, and where the exchange $W_x^{KS}(\mathbf{r})$ and correlation $W_c^{KS}(\mathbf{r})$ potentials are respectively the work done in the fields $\mathscr{E}_x^{KS}(\mathbf{r})$ and $\mathscr{E}_c^{KS}(\mathbf{r})$, which in turn arise from the Kohn–Sham Fermi $\rho_x^{KS}(\mathbf{r}, \mathbf{r}')$ and Coulomb $\rho_c^{KS}(\mathbf{r}, \mathbf{r}')$ holes. Thus

$$W_x^{KS}(\mathbf{r}) = -\int_\infty^\mathbf{r} \mathscr{E}_x^{KS}(\mathbf{r}') \cdot d\mathbf{l}' \quad \text{where} \quad \mathscr{E}_x^{KS}(\mathbf{r}) = \int \frac{\rho_x^{KS}(\mathbf{r}, \mathbf{r}')(\mathbf{r}, \mathbf{r}')}{|\mathbf{r} - \mathbf{r}'|^3} d\mathbf{r}' \tag{81}$$

and

$$W_c^{KS}(\mathbf{r}) = -\int_\infty^\mathbf{r} \mathscr{E}_c^{KS}(\mathbf{r}') \cdot d\mathbf{l}' \quad \text{where} \quad \mathscr{E}_c^{KS}(\mathbf{r}) = \int \frac{\rho_c^{KS}(\mathbf{r}, \mathbf{r}')(\mathbf{r} - \mathbf{r}')}{|\mathbf{r} - \mathbf{r}'|^3} d\mathbf{r}' \tag{82}$$

Furthermore

$$\nabla \times [\mathbf{Z}_{t_c}(\mathbf{r}) + \mathscr{E}_x^{KS}(\mathbf{r}) + \mathscr{E}_c^{KS}(\mathbf{r})] = 0 \tag{83}$$

so that $v_{xc}^{KS}(\mathbf{r})$ is path-independent. Again, for those systems with a certain symmetry such as spherically symmetric atoms, etc., $\nabla \times \mathscr{E}_x^{KS}(\mathbf{r}) = \nabla \times \mathscr{E}_c^{KS}(\mathbf{r}) = 0$, so that the work $W_x^{KS}(\mathbf{r})$ and $W_c^{KS}(\mathbf{r})$ are separately path-independent.

The exchange potential $W_x^{KS}(\mathbf{r})$ can be shown analytically [9] to satisfy the sum rule [20, 22] relating the exchange energy $E_x^{KS}[\rho]$ to its functional derivative $v_x^{KS}(\mathbf{r})$ which is

$$E_x^{KS}[\rho] + \int d\mathbf{r}\rho(\mathbf{r})\mathbf{r} \cdot \nabla v_x^{KS}(\mathbf{r}) = 0 \tag{84}$$

as well as the scaling condition [20, 22]

$$v_x^{KS}(\mathbf{r}; [\rho]) = \lambda v_x^{KS}(\lambda\mathbf{r}; [\rho]) \tag{85}$$

where $\rho_\lambda(\mathbf{r}) = \lambda^3\rho(\lambda\mathbf{r})$. (We note that the satisfaction of both these conditions by $W_x^{KS}(\mathbf{r})$ is independent of the spin-orbitals employed.) Since the dependence of the idempotent density matrix $\gamma_s(\mathbf{r}, \mathbf{r}')$ on the density is unknown, it cannot be shown analytically whether or not $W_x^{KS}(r)$ satisfies the second derivative condition [22]

$$\frac{\delta v_x^{KS}(r)}{\delta \rho(r')} = \frac{\delta v_x^{KS}(r')}{\delta \rho(r)} \tag{86}$$

which is one of symmetry in an interchange of $r$ and $r'$.

In spite of the satisfaction of the sum rule and scaling condition, the work $W_x^{KS}(r)$ is not the functional derivative $v_x^{KS}(r)$. In recent work [23], it has been shown that additional work $\Delta W_x^{KS}(r)$ must be done to obtain $v_x^{KS}(r)$. The work $\Delta W_x^{KS}(r)$ is

$$\Delta W_x^{KS}(r) = -\int_\infty^r y(r') \cdot dl' \tag{87}$$

where the field $y(r) = -z(r; [\gamma_1^c])/\rho(r)$, and $\gamma_1^c(r, r')$ in the first-order correction to the idempotent density-matrix $\gamma_s(r, r')$ in an expansion (in terms of the electron-interaction coupling constant $\lambda$) of the density-matrix $\gamma(r, r')$. Due to the fact that $W_x^{KS}(r)$ satisfies the sum rule of Eq. (84) the field $y(r)$ is such that

$$\int dr \rho(r) r \cdot y(r) = 0 . \tag{88}$$

From a quantum-mechanical perspective, the existence of the field $y(r)$ implies that the corresponding approximate wavefunction incorporates Coulomb correlations. However, if the system wavefunction is a Slater determinant of spin-orbitals, and if the orbitals of the local potential which lead to the same system density are equivalent, then $\gamma(r, r') = \gamma_s(r, r')$ and the field $Z_{t_c}(r) = 0$. Under these conditions, the functional derivative $v_{xc}^{KS}(r) = v_x^{KS}(r) = W_x^{KS}(r)$. On the other hand, consider the case [24] where a local potential generates a density equivalent to the Hartree-Fock theory [25] density. Since the orbitals generated by this local potential are not the same as the Hartee-Fock theory orbitals, the corresponding idempotent density matrices will not be the same, and therefore the field $Z_{t_c}(r)$ in this instance will be finite. We note further that for slowly varying densities, for which the local density approximation (LDA) expression for the exchange energy $E_x^{LDA}[\rho]$ is valid, the functional derivative $v_x^{LDA}(r) = \delta E_x^{LDA}[\rho]/\delta \rho(r)$ and the work $W_x^{KS}(r)$ can also be shown [26, 27] analytically to be equivalent. The work interpretation of this functional derivative and energy in turn leads to a more fundamental understanding [13, 27] of electron correlations within this approximation for exchange and correlation as explained in Sect. 6.3.

### 4.3 Kohn–Sham Theory Energy Components in Terms of Fields

In addition to the interpretation of the energy components $E_H[\rho]$, $E_x^{KS}[\rho]$ and $E_c^{KS}[\rho]$ as being the energy of interaction between the density $\rho(r)$ and the corresponding source charge distributions $\rho(\mathbf{r}')$, $\rho_x^{KS}(\mathbf{r}, \mathbf{r}')$ and $\rho_c^{KS}(\mathbf{r}, \mathbf{r}')$ respectively, it is also possible to write these energy functionals including $T_c[\rho]$ in terms of their source fields $\mathscr{E}_H(\mathbf{r})$, $\mathscr{E}_x^{KS}(\mathbf{r})$, $\mathscr{E}_c^{KS}(\mathbf{r})$ and $Z_{t_c}(\mathbf{r})$. Employing the symmetry in the interchange of $\mathbf{r}$ and $\mathbf{r}'$ of the pair-correlation function $h(\mathbf{r}, \mathbf{r}') = g(\mathbf{r}, \mathbf{r}')/\rho(\mathbf{r}')$, and the relation $\nabla v_{xc}^{KS}(\mathbf{r}) = -[\mathscr{E}_{xc}(\mathbf{r}) + Z_{t_c}(\mathbf{r})]$ in the sum rule of Eq. 56 leads to the following expressions:

$$E_H[\rho] = \int d\mathbf{r}\rho(\mathbf{r})\mathbf{r}\cdot\mathscr{E}_H(\mathbf{r}) \tag{89}$$

$$E_x^{KS}[\rho] = \int d\mathbf{r}\rho(\mathbf{r})\mathbf{r}\cdot\mathscr{E}_x^{KS}(\mathbf{r}) \tag{90}$$

$$E_c^{KS}[\rho] = U_c[\rho] + T_c[\rho] \tag{91}$$

$$U_c[\rho] = \int d\mathbf{r}\rho(\mathbf{r})\ \mathbf{r}\cdot\mathscr{E}_c^{KS}(\mathbf{r}) \tag{92}$$

$$T_c[\rho] = \frac{1}{2}\int d\mathbf{r}\rho(\mathbf{r})\ \mathbf{r}\cdot Z_{t_c}(\mathbf{r})\ . \tag{93}$$

Note that the Hartree $E_H[\rho]$, exchange $E_x^{KS}[\rho]$ and the purely Coulomb correlation $U_c[\rho]$ energies arise from fields derived by Coulomb's law, whereas $T_c[\rho]$ is due to a field that is derived from the kinetic-energy-density tensor.

# 5 Asymptotic Structure of the Kohn–Sham Exchange-Correlation Potential $v_{xc}^{KS}(\mathbf{r})$

In this section we discuss the asymptotic structure of the Kohn–Sham exchange-correlation potential $v_{xc}^{KS}(\mathbf{r})$ and its components from the work perspective (see Eq. (80)) for finite and extended systems, and then present results of application to atoms based on this understanding.

## 5.1 Asymptotic Structure for Finite and Extended Systems

For finite systems such as atoms, it is established [5–8] that the structure $v_{xc}^{KS}(\mathbf{r})$ in the classifically forbidden region is

$$v_{xc}^{KS}(\mathbf{r}) = -\frac{1}{r} - \frac{\alpha}{2r^4} \tag{94}$$

where $\alpha$ is the polarizability of the positive ion. The leading term is the exchange $v_x^{KS}(\mathbf{r})$, and the second term the correlation $v_c^{KS}(\mathbf{r})$ potential inclusive of the correlation-kinetic-energy contribution. The fact that the asymptotic structure is entirely due to Pauli correlation effects is readily explained [9] by the work interpretation, and can be shown to be the structure of the exchange potential $W_x^{KS}(\mathbf{r})$. The work interpretation, however, further allows for the determination of the purely Coulomb correlation $W_c^{KS}(\mathbf{r})$ and correlation-kinetic-energy $W_{t_c}(\mathbf{r})$ components of the correlation potential.

Since the total charge of the Coulomb hole $\rho_c^{KS}(\mathbf{r}, \mathbf{r}')$ is zero, the force field $\mathscr{E}_c^{KS}(\mathbf{r})$ due to this charge and consequently the work done $W_c^{KS}(\mathbf{r})$ to move an electron in this field for asymptotic positions of the electron far from this charge, vanishes. Furthermore, the field $\mathscr{E}_c^{KS}(\mathbf{r})$, and thereby the potential $W_c^{KS}(\mathbf{r})$, vanish more rapidly than their exchange counterparts since [28] the part of the Coulomb hole that is large in magnitude and localized about the nucleus changes sign from negative to positive as the electron position is varied from the deep interior of the atom to asymptotic positions in the classically forbidden region. Thus for finite systems in which the Coulomb hole is localized about the atom or molecule,

$$\mathscr{E}_c^{KS}(\mathbf{r}) \xrightarrow[r \to \infty]{} 0 \qquad (95)$$

and therefore

$$W_c^{KS}(\mathbf{r}) \xrightarrow[r \to \infty]{} 0 \ . \qquad (96)$$

For the asymptotic structure of the field $\mathbf{Z}_{t_c}(\mathbf{r})$, we note that it is proportional to the *difference* of fields $\mathbf{z}(\mathbf{r})$ which depend upon the noninteracting and interacting system density matrices $\gamma_s(\mathbf{r}, \mathbf{r}')$ and $\gamma(\mathbf{r}, \mathbf{r}')$, respectively, through the kinetic-energy-density tensor (see Eqs. (54) and (55)). However, asymptotically as the two co-ordinates coalesce, the density matrices are equivalent [2, 29, 30]:

$$\lim_{\mathbf{r}', \mathbf{r}'' \to \infty} \gamma_s(\mathbf{r}', \mathbf{r}'')|_{\mathbf{r}' = \mathbf{r}'' = \mathbf{r}} = \lim_{\mathbf{r}', \mathbf{r}'' \to \infty} \gamma(\mathbf{r}', \mathbf{r}'')|_{\mathbf{r}' = \mathbf{r}'' = \mathbf{r}}. \qquad (97)$$

Therefore, for asymptotic positions of the electron, the field $\mathbf{Z}_{t_c}(\mathbf{r})$ and the work $W_{t_c}(\mathbf{r})$ to move an electron in the field vanish:

$$\mathbf{Z}_{t_c}(\mathbf{r}) \xrightarrow[r \to \infty]{} 0 \qquad (98)$$

and

$$W_{t_c}(\mathbf{r}) \xrightarrow[r \to \infty]{} 0 \ . \qquad (99)$$

The precise analytical asymptotic structure of the potentials $W_c^{KS}(\mathbf{r})$ and $W_{t_c}(\mathbf{r})$ are at present unknown and under investigation. Numerical studies [28], however, show them to decay more rapidly than the exchange potential.

For finite systems, the Fermi hole $\rho_x^{KS}(\mathbf{r}, \mathbf{r}')$ which has a total charge of (negative) unity, is also localized about the system. Thus, for asymptotic electron positions, the field $\mathscr{E}_x^{KS}(\mathbf{r})$ due to the Fermi hole behaves as

$$\mathscr{E}_x^{KS}(\mathbf{r}) \underset{r \to \infty}{\sim} -\frac{1}{r^2} \qquad (100)$$

and the work $W_x^{KS}(\mathbf{r})$, to which the Kohn–Sham potential $v_{xc}^{KS}(\mathbf{r})$ has now reduced, has the structure

$$v_{xc}^{KS}(\mathbf{r}) \underset{r \to \infty}{=} W_x^{KS}(\mathbf{r}) \sim -\frac{1}{r}. \tag{101}$$

Therefore, for all finite systems, the asymptotic behavior of $v_{xc}^{KS}(\mathbf{r})$ arises from the Fermi hole charge distribution $\rho_x^{KS}(\mathbf{r}, \mathbf{r}')$ and is given *exactly* by the structure of $W_x^{KS}(\mathbf{r})$. The above analysis and conclusions are borne out as shown in the example of the Helium ground-state discussed in Sect. 5.2.1.

There are yet other important consequences of the above conclusion. For asymptotic positions of the electron, the Kohn–Sham differential equation of Eq. (31) reduces to

$$\left[ -\frac{1}{2}\nabla^2 + v(\mathbf{r}) + W_H(\mathbf{r}) + W_x^{KS}(\mathbf{r}) \right]\phi_i(\mathbf{r}) = \varepsilon_i\phi_i(x); \quad i = 1, \dots N. \tag{102}$$

(The work $W_H(\mathbf{r})$ is retained in the equation to ensure there is no self-interaction). In contrast to the Kohn–Sham equation, this differential equation can in practice be solved because the dependence of the Fermi hole $\rho_x^{KS}(\mathbf{r}, \mathbf{r}')$, and thus of the work $W_x^{KS}(\mathbf{r})$, on the orbitals is known. Furthermore, since the solution of this equation leads to the exact asymptotic structure of $v_{xc}^{KS}(\mathbf{r})$, and the fact that Coulomb correlation effects are generally small for finite systems, the highest occupied eigenvalue should approximate well the exact (nonrelativistic) removal energy. This conclusion too is borne out by results given in Sect. 5.2.2.

As shown above, the differential equation of Eq. (102) is the Kohn–Sham equation for asymptotic positions of the electron. However, this is also the differential equation when all Coulomb correlations are neglected, i.e. when the fields $\mathcal{E}_c(\mathbf{r})$ and $Z_{t_c}(\mathbf{r})$ vanish. (Recall that the difference between the work $W_x^{KS}(\mathbf{r})$ and the functional derivative $v_x^{KS}(\mathbf{r})$ arises from Coulomb correlations.) We therefore refer to this equation as being that of the *Work-interpretation Pauli-correlated* approximation. It is also the differential equation originally proposed by Harbola and Sahni [9] for the case when only Pauli-correlations are considered to be present.

For the nonuniform electron density system at a jellium-metal surface, it is generally accepted [5–7,9,31–33] that the asymptotic structure of the Kohn–Sham exchange-correlation potential is the image potential: $v_{xc}^{KS}(x) \sim -1/4x$. In recent work [10] the analytical asymptotic structure of the exchange potential $v_x^{KS}(\mathbf{r})$ has been determined to be

$$v_x^{KS}(\mathbf{r}) = -\frac{\alpha_s(\beta)/2}{x}; \quad \alpha_s(\beta) = \frac{\beta^2 - 1}{\beta^2}\left[ 1 - \frac{\ln(\beta^2 - 1)}{\pi\sqrt{\beta^2 - 1}} \right] \tag{103}$$

where $\beta = \sqrt{W/\varepsilon_F}$, $W$ is the metal surface barrier height, and $\varepsilon_F$ is the Fermi energy. The asymptotic structure of the correlation potential is then $v_c^{KS}(\mathbf{r}) =$

$- [1 - 2\alpha_s(\beta)]/4x$. For metallic densities $\alpha_s(\beta)/2$ ranges from 0.20 to 0.27. For $\beta = \sqrt{2}$, the coefficient is exactly $1/4$ and corresponds to a Wigner–Seitz radius of $r_s \sim 4.1$ which is that of stable jellium. Thus the $-x^{-1}$ dependence of the correlation potential is weak. The fact that the image potential structure of $v_{xc}^{KS}(r)$ is independent of the metal parameters is because [10] the Fermi hole delocalized [34, 35] within the metal is screened by that part of the Coulomb hole which is delocalized, the image structure then arising from the part of the Coulomb hole localized to the surface region. The analytical asymptotic structure of the work $W_x^{KS}(r)$ as derived from the delocalized Fermi hole is presently being determined [36]. This should then shed light on the asymptotic structure of the potentials $W_c^{KS}(r)$ and $W_{t_c}(r)$ at a metal surface. However, the asymptotic structure of $W_x^{KS}(r)$ has been determined [31] numerically for high density metals employing model potential orbitals, and the result shown to be $\simeq -1/4x$.

## 5.2 Results of Application to Atoms

### 5.2.1 Structure of the Exchange, Correlation and Correlation-Kinetic-Energy Fields and Potentials for the Helium Ground State

The work interpretation of Kohn–Sham theory is in terms of the wavefunction $\Psi(x_1, \ldots x_N)$ and the Kohn–Sham spin-orbitals $\phi_i(x)$. The structure of the exchange, correlation and correlation-kinetic-energy components of the fields and potentials are as such most readily determined for the He atom ground-state, since by the choice of an accurate wavefunction $\Psi$, the Kohn–Sham orbitals are simultaneously also known as $\phi_i(x) = [\rho(r)/2]^{1/2}$. The results [28] given in this section are those obtained for the accurate analytical 39-parameter correlated wavefunction of Kinoshita [37].

In Fig. 1 the exchange $\mathscr{E}_x^{KS}(r)$, correlation $\mathscr{E}_c^{KS}(r)$ and exchange-correlation $\mathscr{E}_{xc}(r)$ fields are plotted together with the function $(-1/r^2)$. Observe that all the force fields vanish at the origin, approaching it linearly. This is due to the fact that, for an electron at the nucleus, the Fermi $\rho_x^{KS}(r, r')$, Coulomb $\rho_c^{KS}(r, r')$ and the Fermi-Coulomb $\rho_{xc}(r, r')$ hole charge distributions are all spherically symmetric about it. As such there is no force field at this position of the electron. The structure of both $\mathscr{E}_x^{KS}(r)$ and $\mathscr{E}_c^{KS}(r)$ are similar, both being negative, with the latter being an order of magnitude smaller. The fact that both the fields are negative is interesting in light of the striking differences in the structure of the corresponding source charge distributions. The Fermi hole is negative for all electron positions, whereas the Coulomb hole is both positive and negative and of substantial magnitude depending upon the electron position. (For details of the structure of these holes and further explanations, we refer the reader to the original literature [28]). Also observe, as noted previously, that the correlation field $\mathscr{E}_c^{KS}(r)$ decays far more rapidly than the exchange field $\mathscr{E}_x^{KS}(r)$, essentially vanishing by $r \sim 4$ a.u. The correlation-kinetic energy field $Z_{t_c}(r)$ plotted in

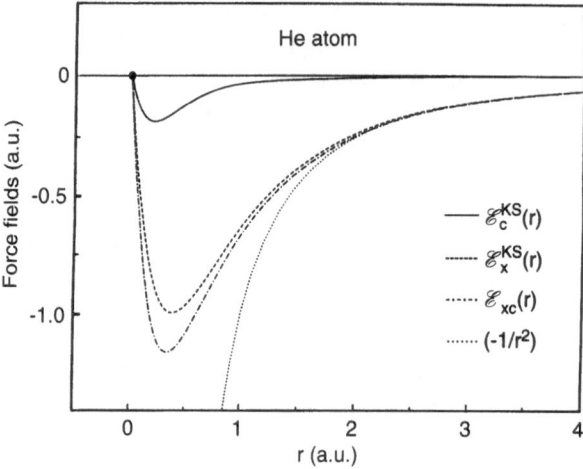

**Fig. 1.** Force fields $\mathscr{E}_x^{KS}(\mathbf{r})$ and $\mathscr{E}_c^{KS}(\mathbf{r})$ due to the Kohn–Sham theory Fermi and Coulomb holes, and the field $\mathscr{E}_{xc}^{KS}(\mathbf{r})$ due to the quantum-mechanical Fermi–Coulomb hole charge distribution for the He atom. The function $(-1/r^2)$ is also plotted

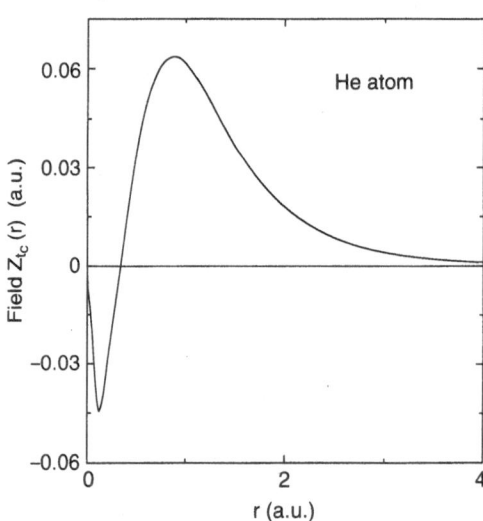

**Fig. 2.** The correlation-kinetic-energy field $Z_{t_c}(\mathbf{r})$ for the He atom

Fig. 2 is also observed to vanish by $r \sim 4$ a.u. It is the exchange field $\mathscr{E}_x^{KS}(\mathbf{r})$ (Fig. 1) that decays asymptotically as $(-1/r^2)$. This then leads to the exchange potential $W_x^{KS}(\mathbf{r})$ having the asymptotic structure $(-1/r)$. In Fig. 3 the exchange $W_x^{KS}(\mathbf{r})$, correlation $W_c^{KS}(\mathbf{r})$ and exchange-correlation $W_{xc}(\mathbf{r})$ potentials are plotted, and in Fig. 4 the correlation-kinetic-energy $W_{t_c}(\mathbf{r})$ and the Kohn–Sham correlation potential (functional derivative) $v_c^{KS}(\mathbf{r})$. Observe that $W_x^{KS}(\mathbf{r})$, $W_c^{KS}(\mathbf{r})$

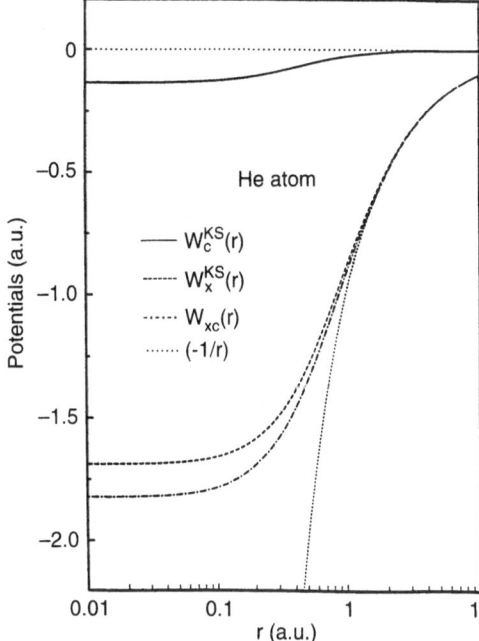

**Fig. 3.** The work-interpretation exchange $W_x^{KS}(\mathbf{r})$, correlation $W_c^{KS}(\mathbf{r})$ and exchange-correlation $W_{xc}(\mathbf{r})$ potentials as determined from the Kohn–Sham theory Fermi and Coulomb holes, and the quantum-mechanical Fermi–Coulomb hole charge, respectively, for the He atom. The fraction $(-1/r)$ is also plotted

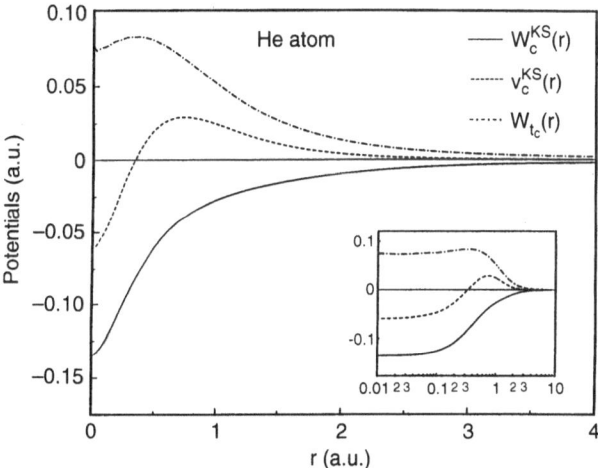

**Fig. 4.** The Work-interpretation correlation $W_c^{KS}(\mathbf{r})$ and correlation-kinetic-energy $W_{t_c}(\mathbf{r})$ potentials for the He atom. The functional derivative $v_c^{KS}(\mathbf{r})$ for the atom is also plotted

and $W_{xc}(\mathbf{r})$ all approach the nucleus quadratically and have zero slope at the origin as they must. In the interior of the atom, the structure of $W_x^{KS}(\mathbf{r})$ and $W_c^{KS}(\mathbf{r})$ are similar, with the latter being an order of magnitude smaller. Both potentials are negative throughout space and monotonic. The potentials $W_c(\mathbf{r})$ and $W_{t_c}(\mathbf{r})$ (see Fig. 4) both vanish by about $r \sim 4$ a.u. so that the asymptotic structure of the Kohn–Sham exchange-correlation potential $v_{xc}^{KS}(\mathbf{r})$ is that of $W_x^{KS}(\mathbf{r}) \sim -1/r$. The potential $W_{t_c}(\mathbf{r})$ is, of course, not monotonic since the field $Z_{t_c}(\mathbf{r})$ changes sign. Furthermore, $W_{t_c}(\mathbf{r})$ also has zero slope at the nucleus since the field $Z_{t_c}(\mathbf{r})$ vanishes there. As is evident (see Fig. 4), the functional derivative $v_c^{KS}(\mathbf{r})$ is also not monotonic. Recent calculations [8] employing a 491-term correlated wavefunction, however, show that $v_c^{KS}(\mathbf{r})$ once again becomes negative for $r > 4$ a.u. and vanishes asymptotically as a negative function. This is consistent with the fact that the polarizability $\alpha$ is positive. The structure of the exchange $W_x^{KS}(\mathbf{r})$ and correlation $W_c^{KS}(\mathbf{r})$ potentials for heavier atoms will be similar and monotonic except that shell structure will be exhibited by a change in slope in the intershell regions. Since the functional derivative $v_c^{KS}(\mathbf{r})$ is known [38] to possess a shell structure (with maxima and minima), the potential $W_{t_c}(\mathbf{r})$ will also exhibit similar nonmonotonicity. However, the asymptotic structure of the functional derivative $v_{ee}^{KS}(\mathbf{r})$ will be that of $W_x^{KS}(\mathbf{r})$ which is $-1/r$.

**Table 1.** Comparison of the highest occupied eigenvalue of the Work-interpretation Pauli-correlated approximation for atoms with those of exact (fully-correlated) Kohn–Sham theory. The negative values in Rydbergs are quoted

| Atom | Work-interpretation Pauli-correlated approximation[a] | Exact[b] | \|difference\| |
|------|------|------|------|
| Atoms with last closed shell on s subshell | | | |
| $^2$He | 1.836 | 1.808 | 0.028 |
| $^4$Be | 0.626 | 0.676 | 0.050 |
| $^{12}$Mg | 0.521 | 0.518 | 0.003 |
| Noble-gas atoms | | | |
| $^{10}$Ne | 1.713 | 1.594 | 0.119 |
| $^{18}$Ar | 1.178 | 1.094 | 0.084 |
| Alkali metals | | | |
| $^3$Li | 0.405 | 0.400 | 0.005 |
| $^{11}$Na | 0.390 | 0.364 | 0.026 |
| Halogens | | | |
| $^9$F | 1.464 | 1.368 | 0.096 |
| $^{17}$Cl | 1.006 | 0.982 | 0.024 |
| Atoms with less than half filled p shells | | | |
| $^5$B | 0.581 | 0.598 | 0.017 |
| $^6$C | 0.818 | 0.820 | 0.002 |
| $^{13}$Al | 0.406 | 0.428 | 0.022 |
| $^{14}$Si | 0.571 | 0.714 | 0.143 |
| Atoms with half and two-thirds filled p shells | | | |
| $^7$N | 1.078 | 1.056 | 0.022 |
| $^8$O | 1.249 | 1.172 | 0.077 |
| $^{15}$P | 0.754 | 0.748 | 0.006 |
| $^{16}$S | 0.861 | 0.832 | 0.029 |

[a] See [41]
[b] See [40]

### 5.2.2 Comparison of Highest-Occupied Eigenvalues of the Work-Interpretation Pauli-Correlated Approximation with Exact Removal Energies

As noted above, the Kohn–Sham equation reduces to the differential equation of Eq. (102) for asymptotic positions of the electron, the structure of the electron-interaction potential there being that of the exchange potential $W_x^{KS}(\mathbf{r})$. Furthermore, Coulomb correlation effects in atoms are small. It is therefore meaningful to compare the highest occupied eigenvalue of this work-interpretation Pauli-correlated approximation differential equation to the exact removal energies. Over the past few years, methods [39] have been developed whereby the Kohn–Sham orbitals and eigenvalues can be determined provided the exact density is assumed known. Thus, with densities determined from configuration-interaction wavefunctions, the highest occupied eigenvalue (and thereby the exact removal energy) for the atoms $^2$He–$^{18}$Ar have recently been obtained [40]. For possible sources of error in the wavefunctions for the heavier atoms and consequently in the Kohn–Sham eigenvalues, we refer the reader to this reference. In Table 1 we compare the highest occupied eigenvalues [41] of Eq. (102) of the work-interpretation Pauli-correlated approximation with those of the Kohn–Sham differential equation. Observe (see last column of Table) that, with the exception of $^{10}$Ne and $^{14}$Si for which the differences are less than two-tenths of a Rydberg, the eigenvalues for the remaining atoms differ by *hundredths or less* of a Rydberg. Thus, accurate removal energies of finite systems can be determined by solution of the differential equation in the work-interpretation Pauli-correlated approximation. For a detailed comparison of the corresponding highest occupied eigenvalues of atoms and negative ions, with experimental ionization potentials and electron affinities respectively, we refer the reader to the literature [41, 42].

### 5.2.3 Excited State Total Energies of the Helium $2^3$S Isoelectronic Sequence

As is known, there is no equivalent of the Hohenberg–Kohn theorems for excited states. However, it was noted in the original work of Harbola and Sahni [9] that an electron in an excited state also has a Fermi–Coulomb hole charge distribution, one that is different from when the electron is in its ground state. Consequently, a local many-body potential can be determined for the excited state as the work done in the field of this charge distribution. Now the physical interpretation of the electron-interaction potential $v_{ee}^{KS}(\mathbf{r})$ in terms of the work done as derived previously via the virial theorem is, of course, rigorous only for the ground-state. On the other hand, the virial theorem (and thus the differential virial theorem) is valid for bound excited states. Further, as noted, the exchange-correlation field $\mathscr{E}_{xc}(\mathbf{r})$ corresponding to an electron in an excited state exists. The structure of a field equivalent to $\mathbf{Z}_{t_c}(\mathbf{r})$ for excited states is unknown, since there is no proof of the existence of a Kohn–Sham differential equation for such states. However, excited states of atoms depend primarily on the asymptotic structure of the effective potential which, once again, must be the exchange potential $W_x^{KS}(r)$ since correlation effects are further diminished for an electron in such a state. As such the excited-state eigenvalues as well as total energies of the work-interpretation Pauli-correlated approximation differential equation of Eq. (102) should again prove to be accurate. This has already been shown [42, 43] to be the case for various excited states of Be and Na. Here we present results of recent calculations on the Helium atom $2^3S$ isoelectronic sequence.

In Table 2 we present numerically refined results [44] of calculations [45] of total energies of the $2^3S$ Helium atom isoelectronic sequence. These energies are determined within the work-interpretation Pauli-correlated approximation from a *single* determinant via solution of Eq. (102). For purposes of comparison we also quote the total energies determined [46] via a 55-term correlated wavefunction calculation. The work-interpretation energies are essentially exact, differing by 0.08 % for He to 0.005 % for $Ne^{8+}$. The increase in accuracy down the isoelectronic sequence is, of course, a consequence of the diminution of Coulomb correlation effects which result because the core orbital shrinks with increasing nuclear charge. Furthermore, since the work-interpretation energies are obtained within the Pauli-correlated approximation, these results show how insignificant Coulomb correlation effects are for this triplet state of the two-electron atom. We expect the same to be true for the excited states of other light-atoms, the results for which may again be obtained by solution of Eq. (102) via a single determinant in the central field approximation [41].

**Table 2.** Total energies of $2^3S$ state of the Helium atom isoelectronic sequence as determined within the Work-interpretation Pauli-correlated approximation, and exact results obtained from correlated wavefunction calculations. The quantities in parenthesis are the percent errors of the Work-interpretation results. The negative values of the energies in Rydbergs are quoted

| Atom/ion | Work-interpretation Pauli-correlated approximation[a] | Exact[b] |
|----------|-------------------------------------------------------|----------|
| He | 4.3470 (0.080) | 4.3505 |
| $Li^+$ | 10.2170 (0.044) | 10.2215 |
| $Be^{2+}$ | 18.5894 (0.026) | 18.5943 |
| $B^{3+}$ | 29.4626 (0.018) | 29.4678 |
| $C^{4+}$ | 42.8361 (0.013) | 42.8415 |
| $N^{5+}$ | 58.7099 (0.009) | 58.7154 |
| $O^{6+}$ | 77.0837 (0.007) | 77.0893 |
| $F^{7+}$ | 97.9576 (0.006) | 97.9632 |
| $Ne^{8+}$ | 121.3316 (0.005) | 121.3373 |

[a] See [44, 45]
[b] See [46]

# 6 Some Consequences of the Work Interpretation of Density-Functional Theory

## 6.1 Comment on the Hohenberg–Kohn Theorem

For the proof of the first part of the Hohenberg–Kohn theorem, according to which the non-degenerate ground-state expectation value of any observable is a unique functional of the density, one needs to establish that two maps C and D are injective (one to one) and thus bijective (fully invertible). Thus, that the inverse maps $C^{-1}$ and $D^{-1}$ exist. These maps are as follows. One defines a set $V$ of local single-particle potentials which lead via solution of the Schrödinger equation (Eq. 5) to a non-degenerate ground-state wavefunction for the N-electron system. The collection of these wavefunctions in the set $\Psi$ defines the map C: $V \rightarrow \Psi$. By construction, each element of $\Psi$ is associated with some element of $V$: the map is surjective. Next, for all ground-state wavefunctions contained in the set $\Psi$ one determines the ground-state density $\rho(\mathbf{r})$ via Eq. (13) establishing the set $\mathscr{P}$ and thereby defining the map D: $\Psi \rightarrow \mathscr{P}$. This map too is surjective. The proof of injectivity of these maps establishes that the inverse maps $C^{-1}$: $\Psi \rightarrow V$ and $D^{-1}$: $\rho(\mathbf{r}) \rightarrow \Psi[\rho]$ exist, which then leads to the statement of the first part of the theorem.

The path whereby the maps C and D are each established is well defined. One solves the Schrödinger equation for each local potential v($\mathbf{r}$) to determine $\Psi$, and then obtains the density $\rho(\mathbf{r})$ from $\Psi$ via its definition. On the other hand, although the inverse maps $C^{-1}$ and $D^{-1}$ are known to exist, the specific paths establishing these maps are thus far unknown. However, the differential form of the virial theorem of Eq. (58) defines the path whereby the external potential v($\mathbf{r}$) is determined from the ground-state wavefunction $\Psi$. The potential v($\mathbf{r}$) is the work done to bring an electron from infinity to its position at $\mathbf{r}$ against the field $\mathbf{F}(\mathbf{r})$ :

$$v(\mathbf{r}) = -\int_{\infty}^{\mathbf{r}} \mathbf{F}(\mathbf{r}') \cdot d\mathbf{l}' \ . \tag{104}$$

The field $\mathbf{F}(\mathbf{r})$ (see Eq. 59) depends on the wavefunction $\Psi$ through the density $\rho(\mathbf{r})$, spinless single-particle density matrix $\gamma(\mathbf{r}, \mathbf{r}')$, and the pair-correlation density g($\mathbf{r}, \mathbf{r}'$). Furthermore, this work is path-independent since the field $\mathbf{F}(\mathbf{r})$ is conservative. The path of the inverse map $C^{-1}$, whereby for every ground-state wavefunction $\Psi$ there corresponds a potential v($\mathbf{r}$), is now well defined.

For degenerate ground-states, each potential $V \in V$ leads to a subspace of wavefunctions $\Psi_V$. Now, since one potential leads to more than one ground-state wavefunction, C as defined previously is no longer a map. However, if V and V' lead to subspaces $\Psi_V$ and $\Psi_{V'}$, and differ by more than a constant, then the inverse map $C^{-1}$: $\Psi \rightarrow V$, where $\Psi$ is a union of the subspaces $\Psi_V$, is a proper map. Certainly ground-state wavefunctions from the subspaces $\Psi_V$ and

$\Psi_{v'}$ will lead to different external potentials via the work done of Eq. (104). As noted previously, if one potential leads to more than one ground-state wavefunction, the relation between potentials and wavefunctions is not a proper map. However, the derivation of the differential form of the virial theorem makes no assumption with regard to the degeneracy of the ground-state wavefunction. It would thus be interesting to learn through Eq. (104) the relationship between the different degenerate states in the subspace $\Psi_V$ and the external potential which gives rise to them.

## 6.2 Understanding of Slater Theory

The precursor to Kohn–Sham density-functional theory is Slater theory [12]. In the latter theory, the nonlocal exchange operator of Hartree–Fock theory [25] is replaced by the Slater local exchange potential $V_x^S(\mathbf{r})$ defined in terms of the Fermi hole $\rho_x(\mathbf{r}, \mathbf{r}')$ as

$$V_x^S(\mathbf{r}) = \int \frac{\rho_x(\mathbf{r}, \mathbf{r}')}{|\mathbf{r} - \mathbf{r}'|} d\mathbf{r}' . \tag{105}$$

The Fermi hole in turn is defined in terms of the idempotent Dirac density matrix $\gamma_s(\mathbf{r}, \mathbf{r}')$ of Eq. (36) where the orbitals $\phi_i(\mathbf{x})$ are solutions of the Hartree–Fock–Slater equation

$$\left[ -\frac{1}{2}\nabla^2 + v(\mathbf{r}) + W_H(\mathbf{r}) + V_x^S(\mathbf{r}) \right] \phi_i(\mathbf{x}) = \varepsilon_i \phi_i(\mathbf{x}). \tag{106}$$

The exchange energy expression, however, remains the same as in Hartree-Fock theory, being the energy of interaction between the density and Fermi hole. From the perspective of the work interpretation, it is evident that the physics underlying the Slater potential $V_x^S(\mathbf{r})$ does not account for the dynamic nature of the Fermi hole charge distribution. The expression for the Slater potential is valid only for static charge distributions. In other words, the effect of the electron on the charge distribution to which it gives rise is not accounted for in Slater theory. As such it does not satisfy the sum rule of Eq. (84) relating the exchange energy to its functional derivative. It does satisfy the scaling condition of Eq. (85). The functional derivative of the Slater potential can also be written [47] as a sum of a local and nonlocal part:

$$\frac{\delta V_x^S(\mathbf{r})}{\delta \rho(\mathbf{r}')} = \delta(\mathbf{r} - \mathbf{r}')\hat{D}(\mathbf{r})V_x^S(\mathbf{r}) + f(\mathbf{r}, \mathbf{r}') \tag{107}$$

where the operator $\hat{D}(\mathbf{r}) = [3\rho(\mathbf{r}) + \mathbf{r}\cdot\nabla\rho(\mathbf{r})]^{-1}(1 + \mathbf{r}\cdot\nabla)$. If this functional derivative is approximated by its local part, then the Slater potential satisfies the [47, 48] second derivative condition of Eq. (86).

As expected in light of the above remarks, the results of Slater theory are not that accurate. Here we compare for atoms the solution [49] of the Hartree–Fock–Slater equation with those of the work-interpretation Pauli-correlated approximation differential equation of Eq. (102). In Fig. 5 we plot the exchange potentials $V_x^S(\mathbf{r})$ and $W_x^{KS}(\mathbf{r})$ for the Argon atom. Observe that, with the exception of the asymptotic region where both $V_x^S(\mathbf{r})$ and $W_x^{KS}(\mathbf{r})$ decay as $(-1/r)$, $V_x^S(\mathbf{r})$ is an underestimate. It is interesting to note that even at $r = 10$ a.u. in the classically forbidden region, the Fermi hole charge distribution is not entirely static. Consequently, even at this electron position, the expression for the Slater potential is not quite exact. In Table 3 we quote the ground state energies of Slater theory [49] and the work-interpretation [41] together with those of Hartree–Fock theory [50]. In Fig. 6 we plot the differences of these energies with respect to Hartree–Fock theory. The Slater theory energies always lie above those of Hartree–Fock, the relative difference between the two diminishing with increasing atomic number. This difference (see Fig. 6) varies from 800 ppm for Be to 64 ppm for Xe. The work interpretation results are superior to those of Slater theory as expected, and also lie above those of Hartree–Fock theory, the corresponding differences, with the exception of Be (137 ppm), being an order of magnitude smaller. From Ne and the heavier atoms, the work interpretation results differ from Hartree–Fock theory by less than 50 ppm, the difference for Xe being 5 ppm. Thus, within the Pauli-correlated approximation, ground-state energies essentially equivalent to those of Hartree–Fock theory are obtained via the local exchange potential of the work interpretation.

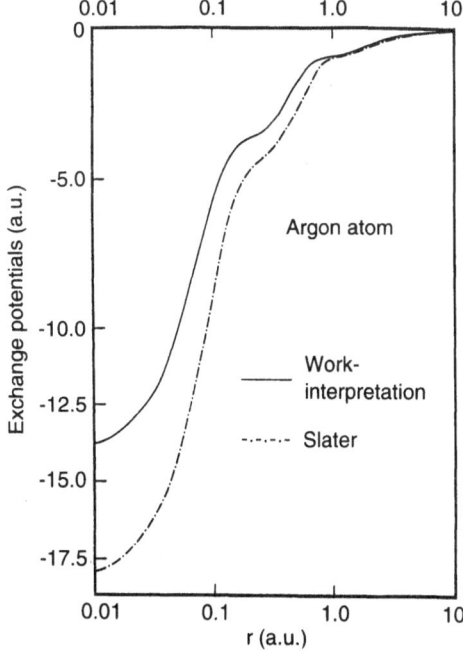

**Fig. 5.** The Slater $V_x^S(\mathbf{r})$ and Work interpretation $W_x(\mathbf{r})$ exchange potentials for the Argon atom

**Table 3.** Total ground-state energies of noble gas and closed s-subshell atoms as determined within Slater theory, the Work-interpretation Pauli-correlated approximation, and Hartree–Fock theory. The negative values of the energies in atomic units are quoted

| Atom | Slater theory[a] | Work-interpretation[b] | Hartree–Fock Theory[c] |
|------|------------------|------------------------|------------------------|
| Be | − 14.561 | − 14.571 | − 14.573 |
| Ne | − 128.501 | − 128.542 | − 128.547 |
| Mg | − 199.533 | − 199.606 | − 199.615 |
| Ar | − 526.703 | − 526.804 | − 526.818 |
| Ca | − 676.606 | − 676.743 | − 676.758 |
| Zn | − 1777.576 | − 1777.820 | − 1777.848 |
| Kr | − 2751.756 | − 2752.030 | − 2752.055 |
| Sr | − 3131.209 | − 3131.519 | − 3131.546 |
| Cd | − 5464.700 | − 5465.093 | − 5465.133 |
| Xe | − 7231.672 | − 7232.101 | − 7232.138 |

[a] See [49]
[b] See [41]
[c] See [50]

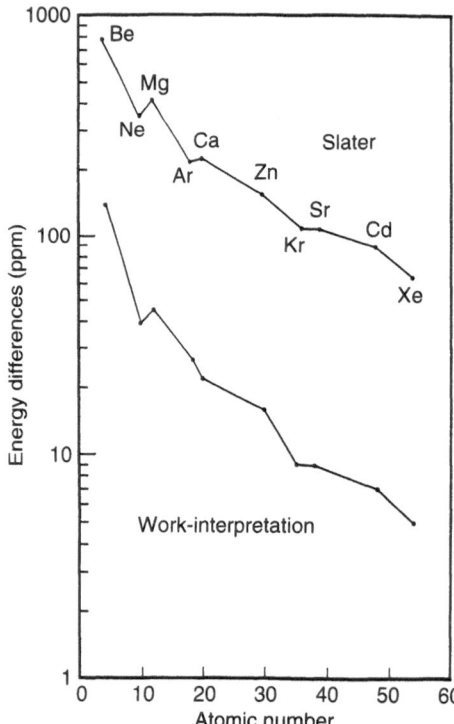

**Fig. 6.** Ground-state energy differences (in ppm) of Slater theory and the Work-interpretation Pauli-correlated approximation with respect to Hartree–Fock theory

For the nonuniform electron gas at a metal surface, the Slater potential has an erroneous asymptotic behavior both in the classically forbidden region as well as in the metal bulk. In the vacuum region, the Slater potential has the analytical [10] asymptotic structure [35, 51] $V_x^S(\mathbf{r}) = -\alpha_S(\beta)/x$, with the coefficient $\alpha_S(\beta)$ defined by Eq. (103). In the metal bulk this potential approaches [35] a value of $(-1)$ in units of $(3k_F/2\pi)$ instead of the correct Kohn–Sham value of $(-2/3)$. Further, in contrast to finite systems, the Slater potential $V_x^S(\mathbf{r})$ and the work $W_x(\mathbf{r})$ are not equivalent [31, 35, 51] asymptotically in the classically forbidden region. This is because, for asymptotic positions of the electron in the vacuum, the Fermi hole continues to spread within the crystal and thus remains a dynamic charge distribution [34].

## 6.3 Electron Correlations in Approximate Kohn–Sham Theory

Since the exchange-correlation energy functional $E_{xc}^{KS}[\rho]$ is unknown, it must be approximated. In approximating this functional, assumptions are made as to how the electrons are correlated, and thus how Pauli and Coulomb correlations are represented in the approximation. The approximate exchange-correlation potential within Kohn–Sham theory is then obtained as the functional derivative of this functional. In following this procedure there is no reason to doubt that the electrons are correlated in a manner other than that originally assumed. However, when the expressions for the approximate potential and energy are rederived via the work interpretation, the existence [27] of electron correlations beyond those assumed in the Kohn–Sham procedure emerge. The work-interpretation derivation shows, furthermore, that both the approximate potential and energy are derivable from the same quantum-mechanical representation of electron correlations, as must be the case. All this allows for a more fundamental understanding of the approximation and of the results obtained therefrom. An analysis of how electrons are correlated within the Kohn–Sham theory, Hartree, Local Density, and Gradient Expansion approximations is given elsewhere [13, 52], and we refer the reader to these papers for details. Here we consider the local density approximation (LDA) for exchange as an example in order to demonstrate the presence of correlations beyond those assumed to exist via the standard Kohn–Sham theory approach to the approximation.

Within the context of Kohn–Sham theory, the assumption underlying the LDA is that each point of the nonuniform electron density is uniform but with a density corresponding to the local value. In the LDA for exchange, the wavefunction is therefore assumed to be a Slater determinant of plane waves at each electron position. The corresponding pair-correlation density $g_x^{(0)}\{\mathbf{r}, \mathbf{r}'; \rho(\mathbf{r})\}$ is thus the expectation of Eq. (66) taken with respect to this Slater determinant, with the resulting expression then assumed valid locally. (The superscript (0) indicates the result is derived from uniform electron gas theory.)

The pair-correlation density thus obtained is

$$g_x^{(0)}\{r, r'; \rho(r)\} = \rho(r') + \rho_x^{(0)}\{r, r'; \rho(r)\} \tag{108}$$

where

$$\rho_x^{(0)}\{r, r'; \rho(r)\} = -\tfrac{1}{2}\rho(r)\left[\frac{9j_1^2(x)}{x^2}\right] \tag{109}$$

is the uniform electron gas Fermi hole, $j_1(x)$ is the first-order spherical Bessel function, $x = k_F R$, $k_F$ is the Fermi momentum, $k_F(r) = [3\pi^2\rho(r)]^{1/3}$, and $R = r' - r$. The Fermi hole $\rho_x^{(0)}\{r, r'; \rho(r)\}$ is by construction *spherically symmetric* about the electron irrespective of its position. The LDA electron-interaction energy $E_{ee}^{(0)}$, which is the energy of interaction between the density and pair-correlation density $g_x^{(0)}\{r, r'; \rho(r)\}$, is then

$$E_{ee}^{(0)} = \frac{1}{2}\int\int\frac{\rho(r)g_x^{(0)}\{r, r'; \rho(r)\}}{|r - r'|}\,dr\,dr'$$

$$= E_H[\rho] + E_x^{LDA}[\rho] \tag{110}$$

where the LDA exchange energy

$$E_x^{LDA}[\rho] = \frac{1}{2}\int\int\frac{\rho(r)\rho_x^{(0)}\{r, r'; \rho(r)\}}{|r - r'|}\,dr\,dr' \tag{111}$$

is the energy of interaction between the density and Fermi hole charge $\rho_x^{(0)}\{r, r'; \rho(r)\}$. The expression for $E_x^{LDA}[\rho]$ can equivalently be written as

$$E_x^{LDA}[\rho] = \int \varepsilon_x^{(0)}\{\rho(r)\}\rho(r)\,dr \tag{112}$$

where $\varepsilon_x^{(0)}\{\rho(r)\} = -3k_F(r)/4\pi$ is the average exchange energy per electron for the uniform electron gas. The electron-interaction potential via Kohn–Sham theory is then

$$v_{ee}^{LDA}(r) = \frac{\delta E_{ee}^{(0)}[\rho]}{\delta\rho(r)} = v_H(r) + v_x^{LDA}(r) \tag{113}$$

where the LDA exchange potential is

$$v_x^{LDA}(r) = \frac{\delta E_x^{LDA}[\rho]}{\delta\rho} = -\frac{k_F(r)}{\pi}. \tag{114}$$

In the work-interpretation derivation, the force field due to the pair-correlation density is first determined, and the potential then obtained as the work done to move an electron in this field. The force field and potential due to $g_x^{(0)}\{r, r'; \rho(r)\}$ are the Hartree field $\mathscr{E}_H(r)$ and potential $W_H(r) = v_H(r)$, respectively, since the spherically symmetric Fermi hole $\rho_x^{(0)}\{r, r'; \rho(r)\}$ does not contribute to the field at the electron position. The field arises only due to the density $\rho(r')$ which is a charge distribution that is *not spherically symmetric* about the electron at $\mathbf{r}$. Thus $g_x^{(0)}\{\mathbf{r}, \mathbf{r}'; \rho(\mathbf{r})\}$ leads via Coulomb's law to an electron-interaction potential that is the Hartree potential $v_H(\mathbf{r})$ rather than $v_{ee}^{LDA}(\mathbf{r})$ of Eq. (113). Therefore it is evident that $g_x^{(0)}\{\mathbf{r}, \mathbf{r}'; \rho(\mathbf{r})\}$ is not the pair-correlation density in the LDA for exchange, and thus not fully representative of how electrons are correlated in this approximation.

The correct LDA pair-correlation density $g_x^{LDA}\{\mathbf{r}, \mathbf{r}'; \rho(\mathbf{r})\}$ is obtained by expanding the general expression for $g_s(\mathbf{r}, \mathbf{r}')$ of Eq. (66) in gradients of the density to $0(\nabla)$ about the uniform electron gas result, and then assuming the resulting expression to be valid locally. The expression thus obtained is

$$g_x^{(0)}\{\mathbf{r}, \mathbf{r}'; \rho(\mathbf{r})\} = \rho(\mathbf{r}') + \rho_x^{(0)}\{\mathbf{r}, \mathbf{r}'; \rho(\mathbf{r})\} + \rho_x^{(1)}\{\mathbf{r}, \mathbf{r}'; \rho(\mathbf{r})\} \qquad (115)$$

where

$$\rho_x^{(1)}\{\mathbf{r}, \mathbf{r}'; \rho(\mathbf{r})\} = \frac{9}{4}\rho(\mathbf{r})\left[\frac{j_0(x)j_1(x)}{k_F^3}\,\hat{\mathbf{R}} \cdot \nabla\, k_F^2\right] \qquad (116)$$

where $j_0(x)$ is the zeroth-order spherical Bessel function, $\hat{\mathbf{R}} = \mathbf{R}/R$. (The superscript (1) indicates the expression to be of $0(\nabla)$.) It is important to note that the lowest order correction term in the expansion for the density $\rho(\mathbf{r})$ is of $0(\nabla^2)$. The term of $0(\nabla)$ is thus a correction to the uniform electron gas Fermi hole. As with the density $\rho(\mathbf{r}')$, the term $\rho_x^{(1)}\{\mathbf{r}, \mathbf{r}'; \rho(\mathbf{r})\}$ is *not spherically symmetric* about the electron and contributes [26] to the force field so that

$$\mathscr{E}^{LDA}(\mathbf{r}) = \int \frac{g_x^{LDA}\{\mathbf{r}, \mathbf{r}'; \rho(\mathbf{r})\}(\mathbf{r} - \mathbf{r}')}{|\mathbf{r} - \mathbf{r}'|^3}\, d\mathbf{r}'$$

$$= \mathscr{E}_H(\mathbf{r}) + \nabla\left[\frac{k_F(\mathbf{r})}{\pi}\right]. \qquad (117)$$

Since $\nabla \times \mathscr{E}_x^{LDA}(\mathbf{r}) = 0$, the work $W_x^{LDA}(\mathbf{r})$ required to move an electron in this field is path independent, and given by

$$W^{LDA}(\mathbf{r}) = W_H(\mathbf{r}) - \frac{k_F(\mathbf{r})}{\pi} \qquad (118)$$

which is the same as Eq. (113) for $v_{ee}^{LDA}(\mathbf{r})$. The electron-interaction energy $E_{ee}^{LDA}$ in turn is the energy of interaction between the density $\rho(\mathbf{r})$ and the pair-correlation density $g_x^{LDA}\{\mathbf{r}, \mathbf{r}'; \rho(\mathbf{r})\}$. However, the non-spherically symmetric component $\rho_x^{(1)}\{\mathbf{r}, \mathbf{r}'; \rho(\mathbf{r})\}$ does not contribute to this integral so that

$$E_{ee}^{LDA} = \frac{1}{2}\int\int \frac{\rho(\mathbf{r})g_x^{LDA}\{\mathbf{r}, \mathbf{r}'; \rho(\mathbf{r})\}}{|\mathbf{r} - \mathbf{r}'|}\, d\mathbf{r}\, d\mathbf{r}'$$

$$= E_H[\rho] + E_x^{LDA}[\rho] \qquad (119)$$

which is the same as Eq. (110) for the electron-interaction energy assumed within Kohn–Sham theory. Thus the work interpretation derivation leads to the same expressions for the potential and energy as derived by the Kohn–Sham scheme. However the derivation shows the existence of additional correlations

33

as represented by the term $\rho_x^{(1)}\{r, r'; \rho(r)\}$. These additional correlations explicitly take into consideration the nonuniformity of the electron density through a term proportional to the gradient of the density. The true Fermi hole in the LDA for exchange is then

$$\rho_x^{LDA}\{r, r'; \rho(r)\} = \rho_x^{(0)}\{r, r'; \rho(r)\} + \rho_x^{(1)}\{r, r'; \rho(r)\} \tag{120}$$

and is a charge distribution that is not spherically symmetric about the electron. Furthermore, it is $\rho_x^{LDA}\{r, r'; \rho(r)\}$ which constitutes the quantum-mechanical source charge distribution giving rise to both the LDA exchange energy and potential. It is obvious from the work-interpretation that in the LDA for correlation, the corresponding Coulomb hole must also contain terms proportional to the gradient of the density, and it is of the form

$$\rho_c^{LDA}\{r, r'; \rho(r)\} = \rho_c^{(0)}\{r, r'; \rho(r)\} + 0(\nabla\rho), \tag{121}$$

where $\rho_c^{(0)}\{r, r'; \rho(r)\}$ is the uniform electron gas Coulomb hole assumed valid locally.

# 7 Conclusions and Future Work

As we have seen, the local electron-interaction potential (functional-derivative) $v_{ee}^{KS}(r)$ of Kohn–Sham density-functional theory in which all the many-body effects are incorporated has a rigorous physical interpretation. It is the work done to move an electron in a field $\mathscr{F}(r)$. The field $\mathscr{F}(r)$, and thus the potential, are comprised of two components. The first is representative of the quantum-mechanical (Pauli and Coulomb) correlations between electrons, the corresponding field $\mathscr{E}_{ee}(r)$ being determined by Coulomb's law. The quantum mechanical source charge distribution for the field $\mathscr{E}_{ee}(r)$ is the pair-correlation density $g(r, r')$. The local potential representing these correlations is then the work done $W_{ee}(r)$ to move an electron in the field $\mathscr{E}_{ee}(r)$. The second component of $\mathscr{F}(r)$ represents the correlation contribution to the kinetic energy. The corresponding field $Z_{t_c}(r)$ is proportional to the difference of fields derived from the kinetic-energy-density tensor $t_{\alpha\beta}(r)$ for the interacting and Kohn–Sham non-interacting systems. The difference between the kinetic-energy-density tensors for these systems may be considered as the 'source' for this field. The local potential representing the correlation contribution to the kinetic energy is then the work done $W_{t_c}(r)$ to move an electron in the field $Z_{t_c}(r)$. Since the field $\mathscr{F}(r)$ is conservative, the work sum $[W_{ee}(r) + W_{t_c}(r)]$ is path-independent. The corresponding quantum-mechanical electron-interaction energy $E_{ee}$ and correlation-kinetic-energy $T_c[\rho]$ can also be derived from their respective sources. The electron-interaction energy is the energy of interaction between the density $\rho(r)$ and the pair-correlation density $g(r, r')$. The correlation-kinetic-energy which is the difference of the interacting and non-interacting kinetic energies is obtained from the difference between the traces of the corresponding kinetic-energy-density tensors.

The physical interpretation of the functional derivative $v_{ee}^{KS}(\mathbf{r})$ as described allows for the *a priori* understanding of its structure and that of its components. For example, the exchange-correlation potential $W_{xc}(\mathbf{r})$, and its exchange $W_x^{KS}(\mathbf{r})$ and correlation $W_c^{KS}(\mathbf{r})$ components all approach the nucleus of an atom quadratically, having zero slope there. This is the case [53] for both spherical and nonspherical density atoms. The reason for this structure is that, for an electron at the nucleus, the Fermi–Coulomb, Fermi and Coulomb hole charge distributions are all spherically-symmetric. As such the corresponding fields at the electron position due to these charge distributions vanish, which in turn leads to the potentials having zero slope there. In the interior of the atom these potentials must exhibit shell structure but be monotonic throughout, since positive work must be done to remove an electron against the force of these fields. Any non-monotonicity of the functional derivative $v_{xc}^{KS}(\mathbf{r})$ can then be attributed to correlation-kinetic-energy effects. Finally, in the classically forbidden region, the asymptotic structure of $v_{ee}^{KS}(\mathbf{r})$ for all finite systems is precisely that of the work $W_x^{KS}(\mathbf{r})$ and can be determined exactly by solution of the differential equation in the work-interpretation Pauli-correlated approximation. The physical interpretation also sheds light on other theories of electronic structure. Thus, for example, the reason why the Slater potential $V_x^S(\mathbf{r})$ approaches the work $W_x^{KS}(\mathbf{r})$ in the classically forbidden region of finite systems is because the Fermi hole becomes an essentially static charge distribution for these electron positions. The fact that the Fermi hole remains a dynamic charge for asymptotic positions of an electron in the vacuum region at a metal-vacuum interface also explains why the two potentials cannot be equivalent in this case.

There is, of course, much that remains to be understood with regard to the physical interpretation. For example, the correlation-kinetic-energy field $Z_{t_c}(\mathbf{r})$ and potential $W_{t_c}(\mathbf{r})$ need to be investigated further. However, since accurate wavefunctions and the Kohn–Sham theory orbitals derived from the resulting density now exist for light atoms [40] and molecules [54], it is possible to determine, as for the Helium atom, the structure of the fields $\mathscr{E}_x^{KS}(\mathbf{r})$, $\mathscr{E}_c^{KS}(\mathbf{r})$, $\mathscr{E}_{xc}(\mathbf{r})$, and $Z_{t_c}(\mathbf{r})$, and the potentials $W_x^{KS}(\mathbf{r})$, $W_c^{KS}(\mathbf{r})$, $W_{xc}(\mathbf{r})$, and $W_{t_c}(\mathbf{r})$ derived from them, respectively. A study of these results should lead to insights into the correlation and correlation-kinetic-energy components, and to the numerical determination of the asymptotic power-law structure of these fields and potentials. The analytical determination of the asymptotic structure of either $[Z_{t_c}(\mathbf{r}), W_{t_c}(\mathbf{r})]$ or $[\mathscr{E}_c^{KS}(\mathbf{r}), W_c^{KS}(\mathbf{r})]$ would then lead to the structure of the other.

There is then the question of understanding the physical origin of the discontinuity [2] of the Kohn–Sham exchange-correlation potential $v_{xc}^{KS}(\mathbf{r})$ as the number N of electrons passes through an integer. It would thus be of interest to learn whether and how each component $W_x^{KS}(\mathbf{r})$, $W_c^{KS}(\mathbf{r})$ and $W_{t_c}(\mathbf{r})$ of the potential contributes to the discontinuity. The addition of an infinitesimal amount of charge changes the density infinitesimally. However, the functional dependence of the pair-correlation density and single-particle density matrix on the density is unknown. Thus, the changes in these properties on addition of a fractional charge, and therefore the change in the respective fields and work done, is also unknown and needs to be investigated.

The physical description of the functional derivative $v_{ee}^{KS}(\mathbf{r})$ requires knowledge of the wavefunction $\Psi$ for the determination of the electron-interaction component $W_{ee}(\mathbf{r}) = W_H(\mathbf{r}) + W_{xc}(\mathbf{r})$, and knowledge of both the wavefunction $\Psi$ and the Kohn–Sham orbitals $\phi_i(\mathbf{x})$ for the correlation-kinetic-energy component $W_{t_c}(\mathbf{r})$. The corresponding Kohn–Sham 'wavefunction' is then a single Slater determinant. It has, however, also been proposed [42, 52, 53] that the wavefunction $\Psi$ be determined by solution of the Sturm-Liouville equation

$$\left[ -\frac{1}{2}\nabla^2 + v(\mathbf{r}) + W_{ee}(\mathbf{r}) \right]\phi_i(\mathbf{x}) = \varepsilon_i\phi_i(\mathbf{x}) \tag{122}$$

where $W_{ee}(\mathbf{r})$ is the work done in the field $\mathscr{E}_{ee}(\mathbf{r})$ of the pair-correlation density $g(\mathbf{r},\mathbf{r}')$. The wavefunction is then of the form

$$\Psi(x_1, \ldots x_N) = \sum_i B_i\Phi_i\{\phi_i(\mathbf{x})\} \tag{123}$$

where the $\Phi_i$ are N-electron determinantal functions formed from the infinite set of spin-orbitals generated by the equation, and the $B_i$ are appropriately chosen coefficients. The construction of an approximate configuration-interaction (CI) wavefunction in this manner differs in fundamental ways from conventional CI calculations. The most significant of these is that the effects of both Pauli and Coulomb correlations are intrinsically incorporated into the structure of the basis functions. This is because the orbitals are generated self-consistently from the field of the pair-correlation density which in turn depends upon the wavefunction. In addition, the asymptotic structure of the orbitals will be correct since the potential $W_{ee}(\mathbf{r})$ decays as $(-1/r)$ in the classically forbidden region. For these reasons it is likely that the number of configurations required to achieve a certain accuracy for the total energy will be reduced from those of standard CI calculations. We reiterate that the description of the physics of electron-interaction whereby the spin-orbitals are generated from a local potential which is the work done in the field of the pair correlation density has already been shown to be accurate at the Pauli-correlated level. Thus, when the wavefunction is a single Slater determinant, ground-state energies of atoms equivalent to those of Hartree–Fock theory are obtained [41]. More recently, we have also shown [55] for atoms that when the wavefunction is assumed to be a product of spin-orbitals, ground-state energies equivalent to those of Hartree theory [55, 56] are derived. For the open-shell atoms, these Hartree, Hartree–Fock and Work-interpretation calculations are all performed in the central-field approximation, so that the work $W_{ee}(\mathbf{r})$ is path-independent. If such an approximation is not made for non-symmetrical density systems, then the curl of the field $\mathscr{E}_{ee}(\mathbf{r})$ may not vanish [57], and the work $W_{ee}(\mathbf{r})$ in the differential equation of Eq. (122) not be path-independent. For such cases an effective electron-interaction potential $W_{ee}^{eff}(r)$ can be constructed from the irrotational component of the field $\mathscr{E}_{ee}(r)$, the solenoidal component being neglected. This is equivalent to determining the potential from an effective *static* electron-interaction charge distribution $\rho_{ee}^{eff}(r)$ so that

$$W_{ee}^{eff}(r) = \int \frac{\rho_{ee}^{eff}(r')}{|r - r'|} dr' \tag{124}$$

where $\rho_{ee}^{eff}(r) = \nabla \cdot \mathscr{E}_{ee}(r)/4\pi$. Calculations [53] for a model nonspherical density atom, for which the curl of the field $\mathscr{E}_{ee}(r)$ does not vanish, show its solenoidal component to be negligible in comparison to the irrotational part, and therefore the corresponding effective potential $W_{ee}^{eff}(r)$ to be accurate. Thus, the use of $W_{ee}^{eff}(r)$ in the differential equation of Eq. (122) should lead to accurate results.

As a consequence of the accuracy of the work-interpretation Pauli-correlated approximation, the exchange-correlation potential $v_{xc}^{KS}(r)$ has recently [58] been approximated by assuming its exchange component to be the work $W_x^{KS}(r)$ whereas its correlation component is derived as the functional derivative of an accurate correlation energy functional constructed [59] by modeling the Coulomb hole. This approximation has led to accurate results for total and removal energies of atoms, and should readily be applicable to molecules and clusters. Finally, we note that there has been other work [60] towards the determination of accurate Coulomb hole charge distributions. In light of the physical interpretation of the Kohn–Sham potential, it is suggested [61] that instead of first constructing an energy functional from these holes and then determining the correlation potential as its functional derivative, the potential be determined instead directly from the hole charge as the work done in its field. This will ensure that the same correlations are assumed in the determination of the energy as well as the potential. Furthermore, since the potential thus derived is determined from a physically correct (albeit approximate) charge distribution which satisfies various constraints, it will not possess any singularities that might exist in the structure of the corresponding functional derivative.

*Acknowledgement.* The author wishes to thank Prof. Roman Nalewajski for his invitation to write this brief review. Thanks are also due to Dr. Manoj Harbola for generating the work-interpretation results of Table 2. Constructive comments by Alexander Solomatin and Prof. Andrzej Holas are also acknowledged.

# 8 References

1. Schrödinger E (1926) Ann Phys 79: 361, 489, 734; 80: 437; 81: 109
2. Parr RG, Yang W (1989) *Density Functional theory of Atoms and Molecules.* Oxford University

Press, Oxford; Dreizler RM, Gross EKU (1990) *Density functional theory.* Springer, Berlin Heidelberg New York; Kryachko ES, Ludeña EV (1990) Energy density functional theory of many-electron systems. Kluwer, Dordrecht; March NH (1992) Electron density theory of atoms and molecules. Academic, London

3. Hohenberg P, Kohn (1964) Phys Rev 136: B864
4. Kohn W, Sham LJ (1965) Phys Rev 140: A1133
5. Williams AR, Barth U von (1983) In: Lundqvist S, March NH (eds) *Theory of the inhomogeneous electron gas.* Plenum, New York
6. Almbladh CO, Barth U von (1985) Phys Rev 31: 3231
7. Sham LJ, (1985) Phys Rev B32: 3876
8. Umrigar CJ, Gonze X (1994) Phys Rev A50: 3827
9. Harbola MK, Sahni V (1989) Phys Rev Lett 62: 489; Sahni V, Harbola MK (1990) Int J Quantum Chem Symp 24: 569
10. Solomatin A, Sahni V (1996) Phys. Lett. A 212: 263
11. Holas A, March NH (1995) Phys Rev A51: 2040
12. Slater JC (1951) Phys Rev 81: 385
13. Sahni V (1996) In: Chong DP (ed) *Recent advances in density functional methods, Part I.* World Scientific
14. Sahni V, Krieger JB (1975) Phys Rev A11: 409; Sahni V, Krieger JB, Gruenebaum J (1975) Phys Rev A12: 768
15. Dirac PAM (1930) Proc Cambridge Philos Soc 26: 376
16. Perdew JP, Parr RG, Levy M, Balduz, JL (1982) Phys Rev Lett 49: 1691; Levy M, Perdew JP, Sahni V (1984) Phys Rev A 30: 2745; Almbladh CO, Barth U von (1985) Phys Rev B31: 3231
17. Nagy A (1990) Phys Rev Lett 65: 2608
18. Harbola MK, Sahni V (1990) Phys Rev Lett 65: 2609
19. Averill FW, Painter GS (1981) Phys Rev B24: 6795
20. Levy M, Perdew JP (1985) Phys Rev A32: 2010
21. Sahni V, Levy M (1986) Phys Rev B33: 3869
22. Ou-Yang H, Levy M (1990) Phys Rev Lett 65: 1036; (1991) Phys Rev A44: 54
23. Levy M, March NH (submitted for publication)
24. Holas A, March NH, Takahashi Y, Zhang C (1993) Phys Rev A48: 2708
25. Fock V, (1930) Z. Physik 61: 126; Z Physik 62: 795; Slater JC (1930) Phys Rev 35: 210
26. Wang Y, Perdew JP, Chevary JA, Macdonald LD, Vosko SH (1990) Phys Rev A41: 78
27. Sahni V (1995) In: Gross EKU, Dreizler RM (eds) *Density-functional theory, Vol. 337 of NATO advanced study institute, series B: Physics.* Plenum, New York; Sahni V, Slamet M (1993) Phys Rev B48: 1910; Slamet M, Sahni V (1992) Phys Rev B45: 4013
28. Slamet M, Sahni V (1995) Phys Rev A51: 2815
29. Tal Y, Bader RFW (1978) Int J Quantum Chem Symp 12: 153; March NH (1981) Phys Lett 84: 319; March NH, Pucci R (1981) J Chem Phys 75: 496
30. Ernzerhof M, Burke K, Perdew JP (submitted for publication)
31. Harbola MK, Sahni V (1989) Phys. Rev. B39: 10437
32. Harbola MK, Sahni V (1993) Int J Quantum Chem Symp 27: 101
33. Equiluz AG, Heinrichsmeier M, Fleszar A, Hanke W (1992) Phys Rev Lett 68: 1359; Equiluz AG, Deisz JJ, Heinrichsmeier M, Fleszar A, Hanke W (1992) Int J Quantum Chem Symp 26: 837
34. Sahni V, Bohnen K-P (1985) Phys Rev B31: 7651; (1984) Phys Rev B29: 1045; Harbola MK, Sahni V (1988) Phys Rev B37: 745
35. Sahni V (1989) Surf Sci 213: 226
36. Solomatin A, Sahni V (unpublished)
37. Kinoshita T (1957) Phys Rev 105: 1490
38. Almbladh CO, Pedroza AC (1984) Phys Rev A29: 2322
39. Jones RS, Trickey SB (1987) Phys Rev B36: 3095 (1987); Wang Y, Parr RG (1993) Phys Rev A47: 1591; Leeuwen R van, Baerends EJ (1994) Phys Rev A49: 2421
40. Morrison RC, Zhao Q (1995) Phys Rev A51: 1980
41. Sahni V, Li Y, Harbola MK (1992) Phys Rev A45: 1434; Li Y, Harbola MK, Krieger JB, Sahni V (1989) Phys Rev A40: 6084
42. Sahni V (1995) In: Calais JL, Kryachko E (eds) *Structure and Dynamics of Atoms and Molecules: Conceptual trends.* Kluwwer, Dordrecht
43. Sen KD (1992) Chem Phys Lett 188: 510

44. Harbola MK (private communication)
45. Singh R, Deb BM (1994) Proc Indian Acad Sci 106: 1321
46. Thakkar AJ, Smith VH Jr (1977) Phys Rev A15: 1; (1977) Phys Rev A15: 16
47. Solomatin A, Sahni V, March NH (1994) Phys Rev B49: 16856
48. Solomatin A, Sahni V (1995) Int J Quantum Chem Symp 29: 31
49. Harbola MK, Sahni V (1993) J Chem Educ 70: 920
50. Fischer CF (1977) *The Hartree-Fock Method for Atoms.* Wiley, New York
51. Harbola MK, Sahni V (1987) Phys Rev B36: 5024
52. Sahni V (1995) Int J Quantum Chem 53: 591
53. Slamet M, Sahni V, Harbola MK (1994) Phys Rev A49: 809; Harbola MK, Slamet M, Sahni V (1991) Phys Lett A157: 60
54. Gritsenko OV, Leeuwen R van, Baerends EJ (1995) Phys Rev A52: 1870
55. Solomatin A, Sen KD, Sahni V (manuscript in preparation)
56. Hartree DR (1928) Proc Cambridge Philos Soc 24: 39; (1928) Proc Cambridge Philos Soc 24: 111; (1928) Proc Cambridge Philos Soc 24: 426
57. Ou-Yang H, Levy M (1990) Phys Rev A41: 4038; Rasolt M, Geldart DJW (1990) Phys Rev Lett 65: 276; Harbola MK, Sahni V (1990) Phys Rev Lett 65: 277
58. Cordero NA, Sen KD, Alonso JA, Balbás LC (1995) J Phys II France 5: 1277
59. Gritsenko OV, Rubio A, Balbás LC, Alonso JA (1993) Phys Rev A47: 1811
60. Perdew JP (1991) In: Ziesche P, Eschrig H (eds) *Electronic Structure of Solids 1991.* Akademie Verlag, Berlin; Perdew JP (1986) Phys Rev B33: 8822
61. Slamet M, Sahni V (1992) Int J Quantum Chem Symp 26: 333

# Application of Density Functional Theory to the Calculation of Force Fields and Vibrational Frequencies of Transition Metal Complexes

Attila Bérces and Tom Ziegler

The University of Calgary, Calgary, Alberta T2N 1N4, Canada

In the last five years we have applied density functional theory to gain information about the harmonic force fields, vibrational frequencies and IR intensities of transition metal complexes. This paper is the summary of the outcome of this series of investigations. We discuss the calculation procedures with special emphasis on the effect of reference geometry and exchange correlation

Topics in Current Chemistry, Vol. 182
© Springer-Verlag Berlin Heidelberg 1996

potential. We also include our benchmark test calculation of the benzene force field. We discuss the major findings of our force field studies of transition metal complexes: ferrocene, debenzene-chromium, benzene-chromium tricarbonyl, and transition metal carbonyls. We found numerous miss assignments in the experimental spectra. We investigated how the force constants of aromatic rings change upon complexation, and we provide explanations for these changes based on qualitative orbital analysis.

# 1 Introduction

The ability to calculate molecular force fields (and geometries) is likely one of the most important developments in computational chemistry over the past twenty years. Until the end of last decade, most force field calculations had been carried out by the Hartree–Fock method. In the last years, *density functional theory* (DF) has emerged as an attractive alternative to traditional *ab initio* techniques. The implementation of analytical energy gradients by Versluis and Ziegler [1], as well as by Fournier et al. [2], has made it possible to evaluate structures and force fields by DF-based methods for a large cross section of systems with considerable success. The implementation of analytic second derivatives within the DF formalism in major computer program systems made the calculation of force constants a routine task [3]. As a consequence, the number of studies on the application of DFT to force-field calculations have increased significantly in the last few years. A recent review paper by Fournier and Pápai [4] contains a detailed literature survey of DFT frequency-and force field calculations.

DF theory has the simplicity of an independent-particle model, yet it can be applied successfully to those systems – such as transition metal complexes – where non-dynamical electron correlation is of primary importance. DF-based methods are, in general, very easy to use, no matter how sophisticated the functional employed to describe the electron correlation. Also, more sophisticated functionals do not increase the computational requirements significantly, as opposed to post Hartree–Fock *ab initio* calculations. The application of approximate density functional theory has been reviewed by Ziegler and others [5].

The application of *ab initio* methods in the calculation of harmonic force fields of transition metal complexes has been hampered by the size of these systems and the need to employ costly post-Hartree–Fock methods, in which electron correlation is taken into account. Thus, the fruitful symbiosis between *ab initio* theory and experiment, to determine empirically scaled quantum mechanical force fields, has been virtually absent in studies of transition metal complexes.

The first DF-based calculations on vibrational frequencies of multi-bonded diatomics [6] (CO, CC, NN) have shown that the shape of the potential surface is predicted very accurately, even for systems where the traditional Hartree–Fock *ab initio* methods were not very successful. The simple DF-based Hartree–Fock–Slater model has been applied to a variety of small main group molecules ($H_2O$, $H_2S$, $NH_3$, $PH_3$, $CH_4$, $SiH_4$, $C_2H_4$, etc.) [7]. These results are in better agreement with experiment than the Hartree–Fock predictions, and the frequencies are usually underestimated, as opposed to the overestimated *ab initio* results. More sophisticated DF methods were first tested by Fan and Ziegler in calculations on the vibrational frequencies and intensities of $Ni(CO)_4$, $Cr(CO)_6$, as well as of some small organic molecules [8].

Since these early studies, many other investigations have confirmed the general success of density functional theory in the prediction of force constants, vibrational frequencies and intensities [8b, 9]. Applications of DFT in force field studies went beyond the validation of the method and revealed many important characteristics of potential energy surfaces which were not accessable by experimental methods. Experimental techniques can provide accurate information about the potential energy surfaces around the equilibrium geometry of small size molecules, by determining the harmonic and unharmonic force constants. However, these techniques become impractical for larger molecules, since the number of harmonic force constants increases quadratically with the number of atoms. Benzene seems to constitute the practical limit in size for the determination of a full harmonic force field. In fact, a system as large as benzene is only tractable because symmetry reduces the number of distinct force constants to 34. It is interesting to note that the sign and magnitude of three of the empirical force constants of benzene are not yet in agreement with theoretical constants, in spite of extensive efforts in the last decade [10].

The information available on harmonic force fields of transition metal compounds is rather limited. Most of these molecules are too large for experimental determination of the complete harmonic force field. For the *logo molecule* of organometallic chemistry, ferrocene, high symmetry reduces the $57 \times 57$ force-constant matrix to 102 independent elements. Dibenzene–chromium and benzene–chromium–tricarbonyl are also highly symmetrical compounds, with 128 and 236 symmetry unique force constants, respectively. These numbers are small compared to the complete set of 2415 and 1326 force constants, respectively. However, in spite of the greatly reduced size of the problem, the observed vibrational frequencies do not provide enough information to determine the complete harmonic force field. Therefore, the empirical determination of the force fields of these molecules usually involve a number of simplifying approximations.

The present paper summarises the findings of our studies of force fields and vibrational frequencies of transition metal complexes. We discuss transition-metal–carbonyl complexes and complexes with small aromatic rings as ligands in detail. Benzene has an important role in this investigation as a ligand, as well as an excellent benchmark test molecule. Accordingly, we also review the findings of our benzene force field in this report.

# 2 Technical Details

## 2.1 The Importance of Reference Geometry and the Exchange Correlation Potential

For density functional methods, different levels of theory are represented by the different approximations of the exchange correlation potential. The two most important approximations are the *local density approximation* (LDA) and the *gradient-corrected*, or *non-local* (NL) density functionals. The former method obtains the expression for the exchange and correlation from the uniform electron gas model, while the latter also takes the first order changes in the density into account. Many of the verification studies have compared the performance of several different functionals for the prediction of the harmonic vibrational frequencies of main group molecules, as well as of transition metal complexes [11]. The general conclusion seems to be that the current approximate density functionals afford quite adequate estimates of frequencies, even at their simplest level, represented by the local density approximation (LDA). Further, it is generally observed that the gradient-corrected (non-local) density functional methods provide more accurate geometries and vibrational frequencies than the simpler local DFT methods. The improved geometry at the higher level of calculation is especially remarkable for the transition metal complexes.

The more accurate frequencies calculated at the non-local level can be explained partly by the improved reference geometry and partly by the contribution of the non-local gradient corrections to the energy Hessian. In a recent study, we looked at these two effects separately. We compared frequencies obtained by the LDA method at various reference geometries with one another, as well as LDA frequencies with LDA/NL ones at given reference points. This study revealed that the choice of reference geometry has a more important role in the outcome of the calculations than the choice of the exchange-correlation functional. We also found that even the simplest LDA method reproduces experimental harmonic frequencies remarkably well, if highly accurate experimental geometries are used as reference points. Here, we provide a brief explanation and summary of these findings.

The importance of reference geometry was pointed out as early as 1966 by Schwendeman, who observed that the calculated frequencies of diatomic molecules improved when they were evaluated at the experimental geometry [12]. Blom and Altona, as well as Pulay and Fogarasi, also suggested that empirically corrected geometries should be used for HF calculations of the force fields and vibrational frequencies of polyatomic molecules [13]. Further, a recent detailed analysis of the effect of the reference geometry on various orders of force constants by Allen and Császár showed that the major part of the error introduced by the erroneous reference geometry is related to the nuclear-nuclear repulsion term [14]. This effect can be demonstrated by simply examining how the electronic and nuclear parts of the total molecular energy contribute to various orders of force constants. Let us separate the electronic and nuclear potential energies as follows:

$$V(q_1, q_2, ..., q_n) = E_e + \sum_{A \neq B} \frac{Z_A Z_B}{|r_A - r_B|} \tag{1}$$

By differentiating Eq. (1) twice with respect to one of the internal stretching coordinates $R_{AB}$, we get:

$$\frac{\partial^2 V(q_1, R_{AB}, ..., q_n)}{\partial R_{AB}^2} = \frac{\partial^2 E_e}{\partial R_{AB}^2} + 2 \frac{Z_A Z_B}{R_{AB}^3} + \cdots \tag{2}$$

It is apparent from Eq. (2), that this stretching force constant is highly sensitive to the value of $R_{AB}$ due to the $1/R_{AB}^3$ dependence of the second term. Therefore, in force-field and frequency calculations the equilibirum ($r_e$) geometries should ideally be used as a reference point, in order to reproduce experimental vibrational frequencies.

The description of electronic structure and the prediction of vibrational frequencies is significantly improved by DF methods compared to HF theory. As the error in the electronic contribution to the force constants become smaller the error in the nuclear term became increasingly dominant. Therefore, since the error in nuclear contribution to the force constants depends only on the reference geometry, one has to pay special attention to this effect when highly accurate force constants are desired. The molecular geometries are also predicted very accurately by DFT methods. However, even a difference as small as 0.001 Å in a reference bond length makes a significant change in the predicted force constants. One cannot expect 0.001 Å accuracy for molecular geometries even from the most sophisticated DFT methods. Although the importance of reference geometry was pointed out in conection with HF calculations [12, 13], this phenomena has never gained much attention, due to the necessity for scaling the HF force constants to reproduce experimental frequencies accurately.

Another reason why optimized geometries are usually used for frequency calculations is of technical nature. This choice greatly simplifies the calculations, since the frequencies can be obtained by a simple diagonalization of the mass-weighted Cartesian force constants. Most major quantum-mechanical programs do not offer other choices of reference geometry. Any geometry other than the optimized geometry would introduce non-zero forces on the atoms at the reference point, and methods that circumvent the non-zero force dilemma are not generally implemented. However, in our frequency calculations at structures other than the optimized geometry, we shall take the non-zero forces into account properly. (See: Computational details.)

Our first example of the effect of reference geometry is the CH-stretching frequencies of benzene, listed in Table 1. The experimental vibrational frequencies [10a] of benzene represent harmonic modes. The CH-stretching frequencies at the LDA reference geometry are underestimated by about 100 cm$^{-1}$. On the

**Table 1.** The CH-stretching vibrational frequencies of benzene calculated at experimental and optimized reference geometries

|  |  |  | LDA | | calculations |
|---|---|---|---|---|---|
|  |  | ref. geom. | $exp_1$[a] | | LDA |
|  |  | CH/Å | 1.084 | | 1.094 |
|  |  | CC/Å | 1.397 | | 1.388 |
|  |  | exp[b] |  | | |
| $A_{1g}$ | 2 | 3191 | | 3193.6 | 3101 |
| $B_{1u}$ | 13 | 3174 | | 3159.8 | 3065 |
| $E_{2g}$ | 7 | 3174 | | 3170.2 | 3075 |
| $E_{1u}$ | 20 | 3181 | | 3184.4 | 3091 |

[1]Frequencies in $cm^{-1}$. [a][64] [b][10a]

**Table 2.** MC-stretching frequencies[1] of chromium- and nickel-carbonyls calculated at different reference geometries

|  |  | LDA | | calculations | |
|---|---|---|---|---|---|
|  | $Cr(CO)_6$ | ref. geom. $exp_1$[b] | | LDA/NL | LDA |
|  | exp[a] | CrC/Å 1.916 | | 1.917 | 1.862 |
| $A_{1g}$ | 379.2 | 383.3 | | 381.2 | 440.1 |
| E | 390.6 | 390.6 | | 388.4 | 447.1 |
| $F_{1u}$ | 668.1[c] | 687.2 | | 683.7 | 747.2 |
| $F_{1u}$ | 440.5[c] | 441.3 | | 438.2 | 504.4 |
|  | $Ni(CO)_4$ |  | | | |
|  | exp[a] | NiC/Å 1.838 | | 1.844 | 1.781 |
| A | 370.8 | 369.1 | | 361.8 | 433.4 |
| $F_2$ | 458.9[c] | 459.8 | | 451.6 | 518.9 |
| $F_2$ | 423.1[c] | 428.1 | | 419.7 | 491.9 |

[1]Frequencies in $cm^{-1}$. [a][65] [b][66] [c]These frequencies contain a significant portion of MC vibrations, but they are not pure MC modes.

other hand, the calculation at the experimental geometry yields excellent agreement between the theoretical and experimental CH-stretching frequencies, with an average deviation of only 5.2 $cm^{-1}$. Although this agreement between experiment and theory is exceptionally good, one has to be careful with its interpretation. This reference geometry does not reflect the equilibrium value. Therefore, to a certain extent, the excellent agreement between theory and experiment is related to cancellation of errors. Nonetheless, most of the 100 $cm^{-1}$ error at the LDA reference geometry can clearly be accounted for by the too long CH bond length.

The next examples are the metal-carbon and carbon-oxygen stretching frequencies of $Cr(CO)_6$ and $Ni(CO)_4$ at experimental, LDA and LDA/NL reference geometries. Table 2 includes the MC-stretching frequencies, and Table 3 includes the CO-stretching frequencies of these systems. The experi-

**Table 3.** CO-stretching frequencies[1] of chromium- and nickel-carbonyls calculated at different reference geometries

| | | LDA | | calculations | |
|---|---|---|---|---|---|
| | $Cr(CO)_6$<br>exp[a] | ref. geom. exp$_1$[b]<br>CO/Å | 1.147 | LDA/NL<br>1.154 | LDA<br>1.147 |
| $A_{1g}$ | 2139.2 | | 2141.4 | 2093.1 | 2159.4 |
| E | 2045.2 | | 2050.1 | 2001.0 | 2065.2 |
| $F_{1u}$ | 2043.7 | | 2031.5 | 1982.0 | 2044.9 |
| | $Ni(CO)_4$<br>exp[a] | CO/Å | 1.141 | 1.150 | 1.143 |
| A | 2154.1 | | 2153.0 | 2086.3 | 2157.6 |
| $F_2$ | 2092.2 | | 2084.9 | 2017.2 | 2088.5 |

[1]Frequencies in $cm^{-1}$. [a] [65] [b] [66]

mental CO frequencies represent harmonic frequencies, while the MC frequencies are the observed fundamental frequencies. It is generally believed that the unharmonicity in the MC frequencies is in the range of a few $cm^{-1}$.

The CrC- and NiC-bond lengths are calculated very accurately by the LDA/NL method. Accordingly, the corresponding MC stretching frequencies are also very accurate at the LDA/NL reference geometry. There is, however, no significant difference between the predictions of CrC- and NiC-stretching frequencies at experimental and LDA/NL reference geometries. The LDA method seriously underestimates the CrC and NiC bond distance; as a result the corresponding CrC- and NiC-stretching frequencies are overestimated by as much as $64 \, cm^{-1}$ (15%).

The CO-stretching frequencies of metal carbonyls are sensitive not only to the CO distance, but also to the metal–carbon distance due to strong coupling. The calculations at experimental geometry yield CO-stretching frequencies in excellent agreement with experiment for both $Cr(CO)_6$ and $Ni(CO)_4$. The average deviations are $10 \, cm^{-1}$ (0.5%) and $4 \, cm^{-1}$ (0.2%), respectively.

The LDA/NL method usually predicts geometries more accurately than the LDA method. However, this is not the case for the CO-bond length of metal carbonyls, where the LDA and experimental values agree very closely. Accordingly, the CO-stretching frequencies calculated at the LDA reference geometry are also quite accurate. However, the metal-carbon bond length, which is seriously underestimated by the LDA method, increases the CO-stretching frequencies slightly through coupling with metal–carbon stretching vibrations. The LDA/NL method yields more accurate MC-bond distances, while it overestimates the CO distance by 0.007 to 0.009 Å. The longer CO bond results in a 50 to $70 \, cm^{-1}$ downshift for the calculated CO stretching frequencies.

The vibrational frequencies selected here are the most prominent examples to demonstrate the importance of reference geometry. In the present paper we omit the deformational frequencies and the CC-stretching frequencies of benzene. The reader should refer to our original paper for a detailed discussion of the complete set of vibrational frequencies [15].

Another choice that has to be made in applied density functional theory is the treatment of the exchange correlation potential. In Table 4, we compare the LDA vibrational frequencies of $Cr(CO)_6$, and $Fe(CO)_5$ with those of the LDA/NL method. This comparison could have been made at any reference geometry; the results would have been similar. We chose the experimental reference geometry. While we have seen significant differences between the geometries at LDA and LDA/NL level, the frequencies in Table 4 do not show such large deviations. The largest deviation in the stretching frequencies is the $10 \text{ cm}^{-1}$ increase of the $F_{1u}$ CrC-stretching frequency of $Cr(CO)_6$ ($687.2 \text{ cm}^{-1}$ by LDA). It is especially remarkable, that the MC-stretching frequencies are not changed significantly by the non-local corrections, even though these have an important effect on the MC bond length. The $E''$ FeCO-bending frequency of $Fe(CO)_5$ increased from 551 to $566 \text{ cm}^{-1}$, which represents the largest deviation for deformational frequencies.

As a conclusion, the gradient-corrected exchange correlation functional affords significant improvement in the geometry compared to the local methods, but has little effect on the calculated force constants. The previous observation, that LDA/NL methods provide better frequencies than LDA methods, is mainly due to the improved reference geometry. Although in most cases the reference

**Table 4.** Comparison of LDA and LDA/NL vibrational frequencies at experimental geometry

| | $Cr(CO)_6$ | | | $Fe(CO)_5$ | |
|---|---|---|---|---|---|
| | LDA | LDA/NL | | LDA | LDA/NL |
| $A_{1g}$ | 2141.4 | 2140.6 | A1' | 2101.4 | 2104.9 |
| | 383.3 | 389.3 | | 2020.2 | 2019.4 |
| E | 2050.1 | 2046.3 | | 453.3 | 455.4 |
| | 390.6 | 393.8 | | 407.7 | 410.4 |
| $F_{1g}$ | 359.3 | 368.4 | A2' | 351.9 | 362.5 |
| $F_{1u}$ | 2031.5 | 2028.1 | E' | 2007.3 | 2008.4 |
| | 687.2 | 697.9 | | 658.0 | 668.2 |
| | 441.3 | 451.6 | | 466.6 | 474.2 |
| | 100.6 | 108.4 | | 425.8 | 435.9 |
| $F_{2g}$ | 535.5 | 548.3 | | 98.9 | 106.4 |
| | 90.7 | 92.8 | | 49.6 | 44.1 |
| $F_{2u}$ | 513.1 | 525.8 | A2" | 2010.0 | 2010.7 |
| | 63.9 | 64.8 | | 616.7 | 627.4 |
| | | | | 486.4 | 493.3 |
| | | | | 101.3 | 110.8 |
| | | | E" | 550.8 | 565.6 |
| | | | | 365.9 | 376.0 |
| | | | | 93.5 | 95.6 |

[1] Frequencies in $\text{cm}^{-1}$.

geometry improves the results obtained with the non-local functional, the CO-bond distance becomes too long, and the corresponding stretching frequencies are too low.

## 2.2 The Representation of Potential Energy Surfaces for Transition Metal Complexes

One of our goals was to compare the force constants of small organic molecules in the their free state and as ligands in transition metal complexes. Since the values of force constants depend on the definition of "internal coordinates", we paid special attention to the representation of force fields. Internal coordinates represent small internal displacements of the atoms of molecule, with the condition that all other internal coordinates remain unchanged upon displacement. Due to this condition, the same internal coordinate, – say a CC stretch – may represent different displacements, depending on the choice of the complimentary coordinates. This problem becomes especially important for transition metal complexes with a multicenter bond. We can illustrate this problem with the example shown on Fig. 1. The arrows represent the direction of displacement of the atoms upon stretching the CC bond. Fig. 1a shows the CC-stretching displacement for a free $C_2H_2$ molecule, while Figs. 1b and 1c each represent that of $C_2H_2$ complexed to a metal $M$. The difference between 1b and 1c is the representation of the skeletal internal coordinates. In Fig. 1b, we show the CC stretching displacement, with the metal–carbon bonds as complementary coordinates. The displacements have to be orthogonal to the metal carbon bond, since that is an internal coordinate here. Another situation is depicted in Fig. 1c; there, the distance between the metal and the centre of the CC bond is defined as the internal coordinate that expresses the metal–carbon stretch. Fig. 1c clearly shows that in this case the displacement is very similar to that in the free molecule. Therefore, this representation is more appropriate for the comparison of force constants.

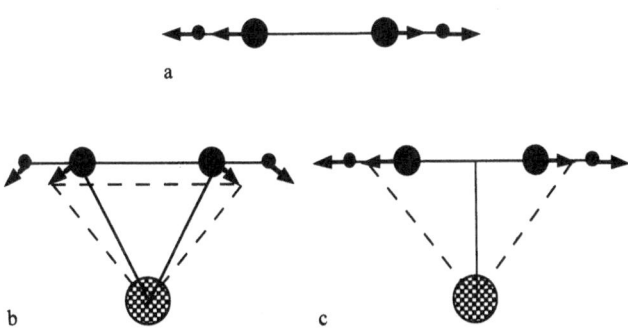

Fig. 1a–c.

The internal coordinates of $\eta^n$-bonded complexes fall naturally into two categories: the ring coordinates and the skeletal modes. We use ferrocene as an example, to demonstrate the skeletal vibrations. The ring coordinates describe distortions of the Cp-ligands whereas the skeletal modes represent movements of the Cp-ligands relative to the metal, without distortion of the Cp-ring. The ring coordinates can be defined following the generally accepted recommendations by Pulay, Fogarasi and co-workers [13b, 13c, 13d], but there is no standard way to select the skeletal internal coordinates.

Before we discuss our definition of skeletal internal coordinates, we rationalize the origin of these degrees of freedom. The extra vibrational degrees of freedom in $FeCp_2$, compared to those of the free Cp-ligands and Fe, are a result

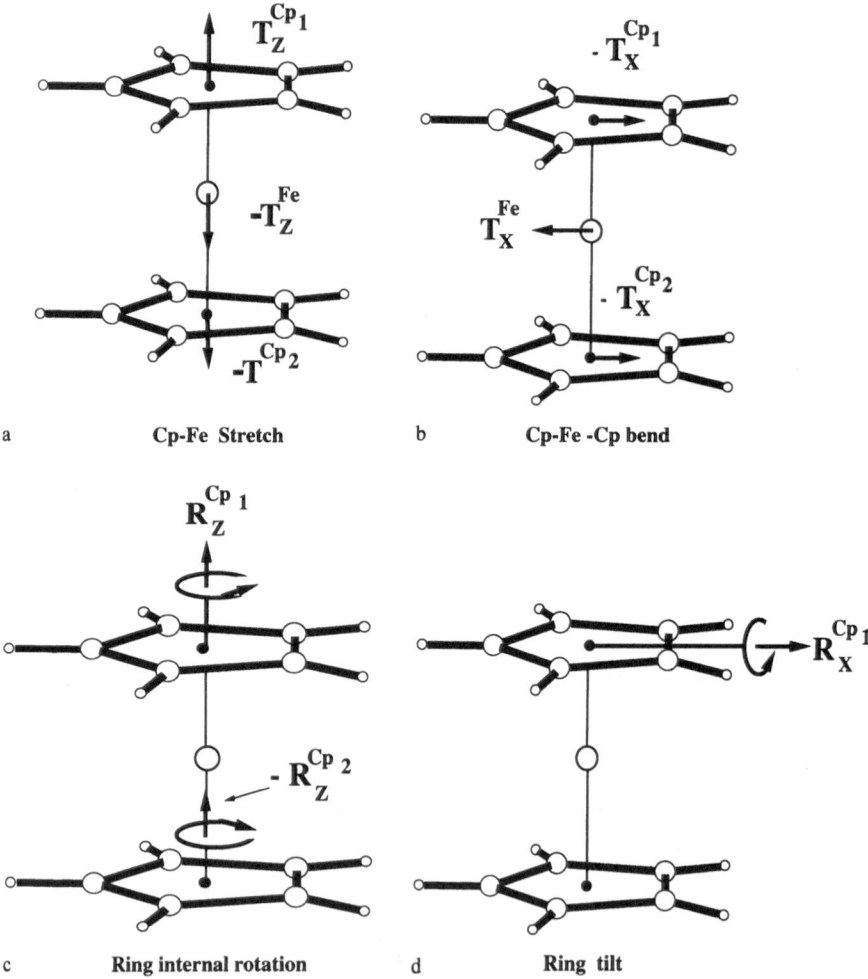

a    Cp-Fe Stretch        b    Cp-Fe -Cp bend

c    Ring internal rotation    d    Ring tilt

**Fig. 2a–d.**

of the interactions between the ligands and the central atom. *This interaction turns the non-vibrational degrees of freedom of the separate entities into vibrational modes*, when they have formed the complex. The six rotational degrees of freedom of the two Cp rings, $R_x^{Cp1}$, $R_x^{Cp2}$ etc., and the nine translational modes of the two rings and the central atom, $T_x^{Cp1}$, $T_x^{Cp2}$, and $T_x^{Fe}$ etc., make up the nine new skeletal vibrations, as well as the six rigid motions of the total complex. The skeletal modes can be separated from the total translations and rotations by taking combinations of the appropriate symmetry.

Structures 2a and 2b demonstrate skeletal metal-ligand stretch and ligand-metal-ligand bending motions, which are related to the translational degrees of freedom of the ligand. Examples of skeletal internal coordinates related to the rotations of the Cp rings are the internal rotation and the ring tilt shown in 2a and 2b.

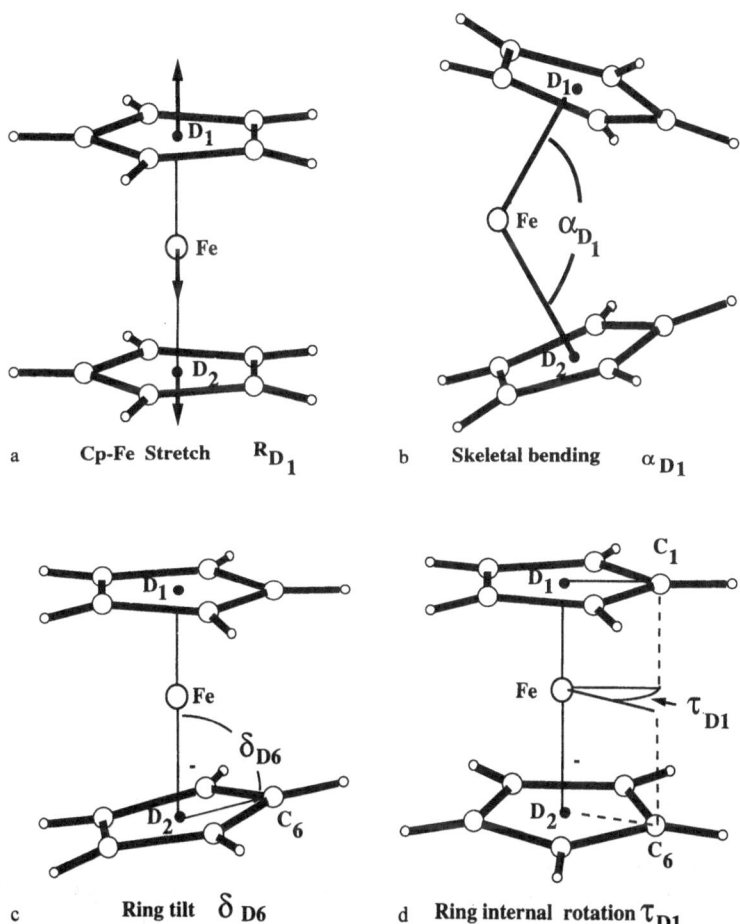

a    Cp-Fe Stretch    $R_{D1}$

b    Skeletal bending    $\alpha_{D1}$

c    Ring tilt    $\delta_{D6}$

d    Ring internal rotation $\tau_{D1}$

Fig. 3a–d.

We studied a few alternative definitions of the skeletal modes, and found that the best representation makes use of a reference point (i.e. dummy atom), that is placed in the geometrical centre of the carbon atoms of the ring. The skeletal stretch can then be represented by the distance between the reference point and the metal atom (3a). The skeletal bend can be represented by the angle between the two metal-reference point vectors (3b).

The tilting coordinate 3c, can be defined by the C–D–Fe angles, and the ring rotation, 3d can be regarded as $C-D_1-D_2-C$ dihedral angle. In order to use a physically meaningful definition for the tilting coordinate, all carbon atoms have to be treated equivalently. This can be done by introducing all possible C–D–Fe angles. This set is a redundant set, since there are only two tilting degrees of freedom for each ring. Redundancies can be eliminated simply by taking symmetrized linear combinations. Even if the molecule has lower symmetry, the highest local symmetry has to be considered in order to find the appropriate internal coordinates. For the internal rotation, the sum of all $C_n-D_1-D_2-C_{n+5}$ $n = 1, ...5$ coordinates are used, to ensure that all carbon atoms are given equivalent consideration.

We note that the Willson type B matrix [16] for the reference point (dummy atom) representation can be evaluated without specific reference to Cartesian coordinates of the dummy atoms, by the chain rule:

$$dq_j(x_1, ..., x_k, x_D(x_1, ..., x_n))/dx_i = dq_j/dx_i + (dq_j/dx_D)(dx_D/dx_i) \qquad (3)$$

where $x_D$ is the dummy atom position, defined as

$$x_D = 1/n \sum_\alpha^n x_a \qquad (4)$$

where $k = 2$ for stretching, $k = 3$ for bending, and $k = 4$ for the torsional internal coordinates. Finally, $n$ is the number of atoms whose geometrical centre defines the position of the dummy atom. Thus, there is *no matrix element between the Cartesian coordinates of the dummy atom and the skeletal internal coordinates*. In this definition of skeletal modes, the dummy atom does not move independently, but only as a result of the displacements of the real atoms of the ring.

## 2.3 Computational Details

The reported calculations were carried out using the Amsterdam Density Functional (ADF) program system developed by Baerends et al. [17] and vectorized by Ravenek [18]. The numerical integration procedure [19] was developed by te Velde et al. A set of auxiliary $s, p, d, f$ and $g$ STO functions [20] – centred on all of the nuclei, was used in order to fit the molecular density and represent the Coulomb and exchange potentials accurately in each SCF cycle. The $1s^2$ configuration on carbon and oxygen as well as the $1s^2 2s^2 2p^6$ configuration of chromium, iron and nickel were assigned to the core and treated by the frozen-core approximation [17]. For transition metals, we used an uncontracted triple-$\zeta$, polarized STO basis set, while a double-$\zeta$, polarized STO basis set was used for C, O, and H [21].

All optimized geometries of transition metal complexes were calculated based on the local density approximation [22] (LDA) energy expression, augmented by nonlocal corrections to exchange [23] and correlation [24]. We shall refer to this method as the LDA/NL scheme, which our group has implemented self-consistently [25a] in energy and gradient [25b] calculations. The optimization of the geometries was based on the GDIIS technique, [26] using natural internal coordinates [13d]. We have interfaced the ADF program with the GDIIS program [27] and implemented the skeletal internal coordinates [28] for geometry optimization. The Cartesian force constants and dipole moment derivatives were calculated by numerical differentiation of the energy gradients [29] and dipole moments using Cartesian displacements. An automatic scheme for the transformation of symmetry-related Cartesian force constants allowed us to make only symmetry-unique displacements [8c]. We used an integration grid with high enough precision to ensure numerical accuracy of $1.0$ cm$^{-1}$, even for the low frequency vibrations.

We have used a locally developed program package [30], based on Schachtschneider's force field program [31] for all force field transformations and the normal coordinate analysis in internal coordinates. For frequencies evaluated at other points than the optimized geometry, the forces or energy gradients have to be taken into account in the force-field transformation [14]. The force constants in internal coordinates can be expressed in terms of the Cartesian energy gradients and energy Hessians as:

$$\frac{\partial^2 V}{\partial q_i \partial q_j} = \sum_{k,l} \frac{\partial^2 V}{\partial x_k \partial x_l} \frac{\partial x_k}{\partial q_i} \frac{\partial x_l}{\partial q_j} + \sum_{l} \frac{\partial V}{\partial x_l} \frac{\partial^2 x_l}{\partial q_i \partial q_j} \tag{5}$$

We have implemented this transformation in our normal coordinate program.

## 3 Verification of the Method by the Reproduction of the Benchmark Benzene Force Field

The harmonic force field of benzene is the most studied force field of any molecule of similar size. Due to the high symmetry of benzene, its harmonic force field can be expressed with only 34 distinct parameters. The modest number of parameters makes it possible to collect enough experimental information to determine the complete force field. Experimentalists have obtained the required data by collecting information on almost all possible isotope-substituted species by various techniques, including non-tranditional vibrational methods based on one-photon and two-photon electron spectroscopy

[32]. The force field of benzene has also been studied by theoretical methods, with the use of extensive basis sets and explicit inclusion of electron correlation [33].

The combination of a Hartree – Fock calculation and experimental information laid the groundwork for the first theoretical force field, due to Pulay et al. [10b]. In this calculation, nine parameters, which incorporated the effect of neglected electron correlation, were fitted to the observed frequencies and Coriolis constants of benzene. The accuracy of the determined fitting parameters was demonstrated by simulating the effect of electron correlation on the calculated HF force field of pyridine [34], naphthaline [35] and other benzene analogs. More elaborate calculations [33c, 33d], including a very recent high level (CCSD(T)) *ab initio* calculations by Zhou et al. [33d] have substantiated the scaled HF force field of Pulay et al.

Through a series of investigations spanning more than a decade, Ozkabak and Goodman (OG) have determined the complete experimental force field of benzene [36]. They base their approach on deriving the force constants from the harmonically corrected experimental frequencies. This procedure has the advantage of producing a better fit of both the CH and CD stretching frequencies, that are differently perturbed by unharmonicities. Further, they collected data

**Table 5.** Calculated and experimental harmonic and fundamental frequencies of benzene ($C_6H_6$)

| | Sym. and Wilson no. | Exp. fundamental[a] /cm$^{-1}$ | estimated harmonic[a]/cm$^{-1}$ | LDA at optimized geom./cm$^{-1}$ |
|---|---|---|---|---|
| $A_{1g}$ | 2 | 3073.942 | 3191 | 3101 |
| | 1 | 993.071 | 994.4 | 1004 |
| $A_{2g}$ | 3 | (1350)[b] | 1367 | 1314 |
| $B_{1u}$ | 13 | (3057)[b] | 3174 | 3065 |
| | 12 | (1010)[b] | 1010 | 993 |
| $B_{2u}$ | 14 | 1309.4 | 1309.4 | 1379 |
| | 15 | 1149.7 | 1149.7 | 1125 |
| $E_{2g}$ | 7 | 3056.7 | 3174 | 3075 |
| | 8 | 1600.9764[c] | 1607 | 1610 |
| | 9 | 1177.776 | 1177.8 | 1150 |
| | 6 | 608.13 | 607.8 | 602 |
| $E_{1u}$ | 20 | 3064.3674[d] | 3181.1 | 3091 |
| | 19 | 1483.9854 | 1494 | 1462 |
| | 18 | 1038.2670 | 1038.3 | 1039 |
| $A_{2u}$ | 11 | 673.97465 | 674.0 | 664 |
| $B_{2g}$ | 5 | (990)[b] | 990 | 985 |
| | 4 | (707)[b] | 707 | 713 |
| $E_{1g}$ | 10 | 847.1 | 847.1 | 830 |
| $E_{2u}$ | 17 | (967)[b] | 967 | 952 |
| | 16 | (398)[b] | 398 | 399 |

[a][10a]. [b]Fundamental frequencies are estimated form infrared combinations. [c]Strong $\nu_8$ and $\nu_1 + \nu_6$ Fermi interactions have been deperturbed. [d]Strong Fermi interactions involving $\nu_{20}$ and three combination bands have been deperturbed.

from the infrared, Raman and one and two-photon electronic spectra of benzene, and of its isotope substituted derivatives. In their most recent paper [10a], they propose these parameters as benchmark values for future theoretical investigations. They have also tested the benchmark force field by simulating eigenvalue- and eigenvector-dependent observables.

The field due to Pulay et al. and the OG field are in good qualitative agreement with each other for most of the symmetry species. Indeed, most of the diffrences are related to the different definitions and assumptions. However, major qualitative differences still remain in the $e_{2g}$ and $e_{1u}$ symmetry species, where one of the $e_{2g}$ force constants is of opposite sign and two of the $e_{1u}$ force constants differ significantly in magnitude.

Handy and co-workers have reported MP2 and Density Functional calculations on the harmonic frequencies of benzene [10e], using an extensive TZ2P $+ f$ basis set. One important outcome of this study is that the density functional calculations could provide very accurate harmonic frequencies at a fraction of the cost of the MP2 calculations.

The valence and symmetry force constants of benzene calculated using density functional theory were first reported by us [10c,d]. These results are summarized in this section. We discuss the vibrational frequencies (Table 5), isotopic shifts, and absorption intensities (Table 6). Selected force constants in symmetry-coordinate representations are listed and compared to the fields due to the Pulay [10b] et al. as well as OG [10a] in Table 7.

### 3.1 $C_6H_6$ Vibrational Frequencies

Table 5 presents frequencies calculated by the Local Density Approximation (LDA) [22] employed in the present study. The calculated frequencies listed in Table 5 are a good demonstration of the predictive power of the LDA method. It follows from Table 5 that calculations at the optimized geometry predict the $C_6H_6$ frequencies with an average deviation of 16.7 cm$^{-1}$ (1.5%), not including CH stretching frequencies.

The CH stretching frequencies ($v_2$, $v_7$, $v_{13}$, $v_{20}$) calculated by the LDA method are close to the experimental fundamental frequencies, but are underestimated by 100 cm$^{-1}$ compared to the harmonic frequencies. The large deviation from the experimental harmonic frequencies can be explained by the sensitivity of calculated CH-stretching frequencies to the applied reference geometry. We have shown in the previous section that the error in the reference bond length is the major source of error in the CH-stretching frequencies.

The largest discrepancy between our prediction and experiment, is the 70 cm$^{-1}$ (5%) overestimate of the $v_{14}$ $b_{2u}$ CC-stretching frequency. This vibrational mode can be characterized as an alternating expansion and contraction of the CC bonds, in which the CC-stretching coupling force constants are of primary importance [10b]. This frequency is underestimated by the HF and

MP2 methods, due to the overestimation of the magnitude of the CC-stretching interaction constants, indicating the preference of these models for localized Kekule structures. Both the HF and MP2 methods are biased towards localized bonds [37], unlike the LDA approach [38], in which such a bias is not present.

This frequency is overestimated in our calculation due to the 5% overestimate of the diagonal and 10–25% underestimate of the off-diagonal CC-stretching constants. We have also calculated the vibrational frequencies at the approximate $r_e$ geometry [10b]. In this set of results, the $v_{14}$ $b_{2u}$ frequency deviates by only 31 cm$^{-1}$; the improvement of this frequency is explained by the better diagonal CC-stretching force constant. The comparison of mode displacements, however, indicates that our force field (at $r_e$) still does not reproduce this vibrational mode correctly. This mode has been studied extensively by Ozkabak et al. [33b]. The empirical field reproduces this mode correctly, as was demonstrated by two-photon crossection calculations. We compared the magnitudes of displacements of the carbon and hydrogen atoms. Note, that all C-atom displacements are equal; the same holds for the H atoms in this symmetry species. The magnitudes of the displacements describe this mode sufficiently well, as all these motions are perpendicular to the CH bond. Also, the H and C-atoms move in opposite directions in Mode 14. The empirical field results in a displacement of 0.106 au. for carbons, which compares well with the value of 0.117 au from calculations based on the LDA force field. However, the calculated displacements of the H atoms are incorrect; their empirical and theoretical values are 0.174 and 0.043 au, respectively. We have tested the mode displacements of $v_{15}$, which is also of $B_{2u}$ symmetry. For $v_{15}$ the displacements based on the empirical and LDA (at $r_e$) force fields were virtually identical.

The $v_6$ and $v_{12}$ ring-deformation frequencies are also somewhat lower than the experimental values. The order of the 1010 cm$^{-1}$ ring bending mode ($v_{12}$) and the 994.4 cm$^{-1}$ CC-stretching frequency ($v_1$) is interchanged in the theoretical spectrum. The inversion is related to the opposite signs of the errors in the CC-stretching and ring-bending force constants. Another discrepancy between the calculated and theoretical frequency order is noticed for the $a_{2g}$ CH planar deformation and the $b_{2u}$ CC-stretching frequencies, which are 1367 and 1309 cm$^{-1}$ experimentally but are 1314 cm$^{-1}$ and 1379 cm$^{-1}$, respectively, based on the LDA calculations.

The out-of-plane force constants are generally an order of magnitude smaller than the in plane constants, and are therefore much more sensitive to numerical errors. The high accuracy employed in the present set of calculations was required for this set of frequencies in particular. The average deviation of the $C_6H_6$ and $C_6D_6$ out-of-plane frequencies from experiment is only 9.4 cm$^{-1}$ (1.2%) and 7.3 cm$^{-1}$ (1.2%), respectively. The frequencies of the $v_{16}$ and $v_4$, torsion modes are most accurate; both agree with the experimental values within 6 cm$^{-1}$. The $v_5$, $v_{10}$, $v_{11}$ and $v_{17}$ CH out-of-plane-deformation modes are slightly underestimated.

## 3.2 Eigenvector-Sensitive Quantities

Ozkabak et al. [33b] demonstrated, in connection with the $b_{2u}$ force constants, that fundamental frequencies provide a poor criterion for the accuracy of a force field. Even good frequency predictions do not necessarily indicate accurate force constants. They suggest that eigenvector-dependent quantities, such as two-photon cross sections, provide a more stringent test.

Infrared absorption intensities are also eigenvector-dependent quantities. However, the accuracy of the calculated absorption intensities depends on the dipole moment derivatives as well. There are only four infrared-active vibrations of benzene. One of these is a pure CH out-of-plane $a_{2u}$ deformation mode, and the corresponding $\nu_{11}$ frequencies for both $C_6H_6$ and $C_6D_6$ are predicted very accurately. The eigenvector corresponding to $\nu_{11}$ is mainly determined by the symmetry of the molecule. Consequently, the error in the calculated intensity is determined by the dipole moment derivatives, rather than by the accuracy of the eigenvector. The LDA-calculated intensity is 85 km/mol, which is in impressive agreement with the experimentally measured absolute intensity of 88 km/mol [39, 40]. Note that the empirical absolute intensities carry about $\pm$ 10% uncertainty, due to the overlapping combination bands and experimental difficulties. The corresponding benchmark- and calculated dipole-moment derivatives are $\partial\mu/\partial S_{11} = 1.285$ D/Å and 1.369 D/Å, respectively. The former, empirical dipole, derivative has been determined from $^{13}C_6H_6$ measurements [40] and reproduces the intensity of about 75 km/mol for both $C_6H_6$ and $^{13}C_6H_6$.

The other three IR-active modes are in plane vibrations of $e_{1u}$ symmetry. Except for the CH-stretching frequency, the error in the $e_{1u}$ frequencies is small; however, it is not systematic for either $C_6H_6$ or $C_6D_6$, indicating that the deformation- and CC-stretching modes are not precisely described in the LDA force field. Accordingly, the in plane absorption intensities are not as accurate as those of the $a_{2u}$ mode. The CH-stretching mode, $\nu_{20}$, has a calculated intensity of 35.8 km/mol, as contracted with the experimental value of 56 km/mol. The calculated intensity of the CC-stretching mode $\nu_{19}$ is 12.2 km/mol, compared to the experimental value of between 14–15 km/mol. The $\nu_{18}$ bending mode intensity is overestimated; the experimental value of 7.5–8.8 km/mol is significantly lower than the calculated estimate of 12.5 km/mol. The dipole-moment derivatives are compared to the empirical values in Table 6. In this table, we also included the calculated intensities for three $D_{6h}$ isotopomers of benzene from different combinations of empirical and DF eigenvectors and dipole moment derivatives. The intensities calculated using empirical dipole-moment derivatives allow us to test the accuracy of the eigenvectors. As it is apparent from these values, the overestimate of $I_{18}$ for $C_6H_6$ is mainly the result of the improper eigenvector. Even with the empirical dipole-moment derivatives it is about the same as $I_{19}$, as opposed to the experimental ratio of about 2 to 3 $I_{20}$ is calculated correctly using empirical tensors. This indicates that the theoretical eigenvector for this CH-stretching mode is correct, principally because there is not much coupling with the CH-bending and CC-stretching vibrations. Modes

**Table 6.** Comparison of calculated intensities using empirical and DFT dipole derivatives and eigenvectors

| | | empirical dipole moment derivatives[a] $\partial\mu/\partial S_{18a} = 0.494,$[b] $\partial\mu/\partial S_{19a} = 0.394,$ $\partial\mu/\partial S_{20a} = 0.770$ | | | DFT dipole moment derivatives $\partial\mu/\partial S_{18a} = 0.533,$[c] $\partial\mu/\partial S_{19a} = 0.478,$ $\partial\mu/\partial S_{20a} = 0.630$ | | |
|---|---|---|---|---|---|---|---|
| | | $C_6H_6,$ | $C_6D_6,$ | $^{13}C_6H_6$ | $C_6H_6,$ | $C_6D_6,$ | $^{13}C_6H_6$ |
| empirical | $I_{18}$ | 7.51 | 7.2 | 6.5 | 9.8 | 9.0 | 8.6 |
| eigenvectors | $I_{19}$ | 11.6 | 2.3 | 12.6 | 13.6 | 2.7 | 14.8 |
| | $I_{20}$ | 56.0 | 30.6 | 55.6 | 37.5 | 20.4 | 37.2 |
| LDA eigenvectors | $I_{18}$ | 10.1 | 8.5 | 8.9 | 12.5 | 10.3 | 11.1 |
| | $I_{19}$ | 10.6 | 1.7 | 11.8 | 12.2 | 2.1 | 13.6 |
| | $I_{20}$ | 54.0 | 29.6 | 53.6 | 35.8 | 19.5 | 35.5 |

[a] Determined from $^{13}C_6H_6$ measurements [40]. Units are D/Å for dipole moment derivatives and km/mol for intensities. Coordinates are defined in [42]. Symmetry coordinates for CH deformation are scaled by $r_{CH}$, so all dipole derivatives are in D/Å. [b][10a], $r_{CH} = 1.084$Å [c]$r_{CH} = 1.094$Å.

18 and 19 both contain significant contributions from CC-stretching and CH-bending coordinates making the eigenvectors more sensitive to small errors in the force constants. The too-high stretching- and too-low bending-force constants present an unfortunate situation for the eigenvectors for these two strongly coupled modes.

It is interesting to compare the intensities calculated using empirical eigenvectors and theoretical polar tensors. As both the theoretical values of $\partial\mu/\partial S_{18a}$ and $\partial\mu/\partial S_{19a}$ are overestimated, the intensity pattern of mode 18 and 19 resembles the empirical pattern, closely, but the absolute values are somewhat too high. On the other hand, the CH-stretching intensity is underestimated. Due to non-systematic errors in the dipole-moment derivatives, the intensity predictions are expected to be only in qualitative agreement with experiment, even when correct eigenvector is used.

Although the frequency shift is an eigenvalue-dependent quantity, it is nevertheless another valid criterion for the quality of a force field and the eigenvectors, provided that the fundamental frequencies are also well reproduced. Accordingly, we have calculated the fundamental frequencies of all $D_{6h}$ isotopomers. We expect that the out-of-plane force field, which predicts the frequencies very accurately, would also give good results for isotopic shifts. The average error in the out of plane $C_6H_6 \rightarrow C_6D_6$ isotopic shifts is 3.5 cm$^{-1}$ (1.6%). The experimental and (calculated) $C_6H_6 \rightarrow C_6D_6$ frequency shifts for the $\nu_{11}, \nu_5, \nu_4, \nu_{10}, \nu_{17}, \nu_{16}$ modes are: $-178$ cm$^{-1}$ ($-177$ cm$^{-1}$), $-161$ cm$^{-1}$ ($-161$ cm$^{-1}$), $-108$ cm$^{-1}$ ($-110$ cm$^{-1}$), $-187$ cm$^{-1}$ ($-184$ cm$^{-1}$), $-180$ cm$^{-1}$ ($-174$ cm$^{-1}$), $-53$ cm$^{-1}$ ($-53$ cm$^{-1}$), respectively. Further evidence for the good quality of our out-of-plane force field is the isotopic shift of the intense $a_{2u}$ infrared-active out-of-plane CH deformation frequency, which is accurately measured by high-resolution interferometric FTIR measurements [41]. The experimental $^{12}C_6H_6 \rightarrow {}^{12}C_6H_5D$ and $^{12}C_6H_6 \rightarrow {}^{13}C^{12}C_5H_6$ shifts are $-67.0$ and $-0.4$ cm$^{-1}$, respectively, both of which are in excellent agreement with values of $-65.3$ and $-0.4$ cm$^{-1}$, respectively, force field predicted by the LDA.

The isotopic shifts of the in-plane frequencies are, in general, not as good as the out-of-plane shifts. Most of the errors are consistent with our expectation, based on the error in the force constants. The $C_6H_6 \rightarrow C_6D_6$ frequency shifts of the CH- and CD-stretching vibrations is an average 29 cm$^{-1}$ (3.6%) smaller than the experimental harmonic shifts. The frequency prediction for these frequencies is improved at the approximate equilibrium reference geometry. Accordingly, the CH- and CD-stretching frequency shifts are closer to the experimental values, these are predicted with an average error of 13.6 cm$^{-1}$ (1.7%). Also, similarly to the frequencies, the frequency shifts are overestimated at this geometry. The CC-stretching interaction constants are underestimated in magnitude. Accordingly, eigenvectors of the CC-stretching modes are not of the same accuracy as eigenvectors of other modes. This problem is most serious in the $b_{2u}$ symmetry. Due to the overestimation of the $b_{2u}$ CC-stretching symmetry force constant and the underestimation of the planar CH-deformation constant, the stretching and deformation modes are more widely separated in the calculated eigenvectors than in the real vibrations. This can be seen from the CC-stretching frequency shift for the $C_6D_6$ isotope, that is 4 cm$^{-1}$ by the calculation, incontrast to 23 cm$^{-1}$, as determined experimentally. On the other hand, the $^{12}C_6H_6 \rightarrow {}^{13}C_6H_6$ isotopic shift for the CH deformation mode is $-10$ cm$^{-1}$ experimentally and only $-2$ cm$^{-1}$ according to our LDA calculation. Interestingly, the $a_{1g}$ CC-stretching frequency shift is always within 1.5 cm$^{-1}$ of the experimental shifts, this is explained by the larger separation of the CH- and CC-stretching modes, as well as by the fact that the CC-stretching coupling constants do not contribute as much to the CC-stretching force constant in the $A_{1g}$ as in the $B_{2u}$ symmetry species.

### 3.3 Force-field Comparison

The empirical force field determination by Ozkabak and Goodman and the scaled *ab initio* filed by Pulay, Fogarasi and Boggs are both based on the most careful approach of its kind. The theoretical and empirical fields, in spite of the very careful studies, are different in magnitude and the sign of some interaction constants. It is not possible to tell which force field is closer to the physical force field of benzene without further studies. One of the objectives of this investigation was to determine whether the differences could be attributed to the assumptions made in the Hartree–Fock calculation. We found, that in spite of the differences in the quantum mechanical models, the scaled Hartree–Fock and density functional force constants are very close in magnitude and agree in sign.

Attila Bérces and Tom Ziegler

Comparison to experimental results is more conveniently carried out in the full symmetry-coordinate representation, as this is the representation in which the experimental force constants are directly determined. Symmetry coordinates are also preferable to a valence-coordinate representation, as any error in the experimental symmetry force constants due to unresolved effects will spread over the entire force field when transformed into the valence-coordinate representation. We have used the full set of symmetry coordinates introduced by Whiffen [42].

The DF calculated symmetry force constants are in very good agreement with the scaled *ab initio* force field [10c, 10d]. Except for the $E_{2g}$ and $E_{1u}$ blocks, the DF force constants are also in excellent qualitative agreement with the empirical constants. Also, the quantitative differences are small and systematic. In the $E_{2g}$ block, the $F_{7,9}$ coupling constant agrees in sign and magnitude with the scaled *ab initio* values, whereas this constant is of opposite sign in the experimental OG field. The other controversial force constants are the $F_{19,20}$ and $F_{18,20}$ interaction force constants in the $E_{1u}$ symmetry block, which are of the same sign in the fields due to Pulay et al. and OG but differ in magnitude. The DF value of these constants is in very good agreement with the scaled *ab initio* results. The $F_{18,20}$ constant is 0.002 mdyn/Å in our calculation, well in line with scaled HF value of 0.006 mdyn/Å, whereas it is 0.151 mdyn/Å in the OG field. As the theoretical constants are so simiar, although they were obtained by two independent quantum chemical approaches, the discrepancy between experiment and theory is more likely related to assumptions in the experimental force-field determination.

As the comparison of theoretical and experimental force constants of benzene shows, it is a very difficult task to determine complete force fields, even for highly symmetrical medium-size molecules. We are primarily interested in the determination of force fields, for substantially larger molecules, like the transition metal complexes of benzene for example. For these larger systems, empirical determination of the force field is impractical, and the theoretical determination of force constants based on density functional theory seems to be a viable method. This application will be discussed in the next sections.

**Table 7.** Disputed symmetry-force constants of benzene. Comparison with the benchmark empirical- and scaled *ab initio* fields[a]

| Sym. | Force constants | Ozkabak-Goodman[d] | LDA optimized geom.[b] | Pulay et al. Set II.[e] |
|------|------|------|------|------|
| $E_{2g}$ | $F_{7,9}(r, \beta)$ | − 0.066 | 0.038 | 0.028 |
| $E_{1u}$ | $F_{18,20}(\beta, r)$ | 0.151 | 0.002 | 0.006 |
| $E_{1u}$ | $F_{19,20}(R,r)$ | 0.572 | 0.186 | 0.175 |

[a] Coordinates are defined in Ref. [42]. Symmetry coordinates for ring deformation and for CH rock and wag are scaled by $r_{CC}$ and $r_{CH}$, respectively, so all force constants are in mdyn/Å. [b] $r_{CC} = 1.388$Å, $r_{CH} = 1.094$Å [d] [10a]; $r_{CC} = 1.397$Å, $r_{CH} = 1.084$Å [e] [10b]; $r_{CC} = 1.395$Å, $r_{CH} = 1.077$Å

# 4 The Vibrational Frequencies and Harmonic Force Fields of Transition Metal Complexes

One objective of the present investigation was to acquire knowledge about how the interaction between the ligand and a metal atom influences the force constants of the ligand. In particular, we are interested in the transition metal complexes of the five and six membered aromatic rings. Having studied the force constants of ferrocene, dibenzene-chromium, and benzene-chormium-tricarbonyl, we compare them with the force fields of the free $Cp^-$ ring, and free benzene. We also make a comparison between analogous force constants of ferrocene and LiCp, as well as of $BzCr(CO)_3$ and $Cr(CO)_6$. We shall, in addition, determine the force constants for the skeletal modes that describe the motion of the ligand(s) relative to the metal centre, and assess the magnitude of the coupling force constants between the skeletal and ring coordinates. The degree to which kinematic coupling effects contribute to the change in the benzene and $Cp^-$ ring-vibrational frequencies that occur on complexation will also be addressed. The accuracy of the calculated force field can be tested by comparing the corresponding frequencies to the experimental ones. This comparison also helps to confirm the assignment of the vibrational spectrum of these systems – or to modify them.

We have seen in the first section how important it is to have an accurate reference geometry for frequency calculations. Therefore, we start with the comparison of empirical and calculated geometries. For the determination of geometries we have used non-local corrections in the exchange correlation potential (LDA/NL). This geometry was used for the determination of LDA force field, and the non-zero forces were taken into account in the calculation of internal force constants and vibrational frequencies. Further, we compare the calculated and observed vibrational frequencies of the transition metal complexes. We also discuss the differences between force constants of free and complexed small aromatic rings.

## 4.1 Geometry and Conformation

Electron-diffraction data in the gas phase [43] suggest that ferrocene prefers to adapt an eclipsed conformation, with an internal rotational barrier of $0.9 \pm 0.3$ kcal/mol. The calculated barrier derived from the vibrational frequency of the internal rotational mode is 0.72 kcal/mol [44]. Our LDA/NL calculation finds the eclipsed conformation to be the most stable, with a calculated rotational barrier of 0.69 kcal/mol, in good agreement with experiment. The structure of ferrocene has been studied by several theoretical methods. Our optimized geometrical parameters are similar to those that had been obtained previously by Fan and Ziegler [25] employing the same LDA/NL scheme (Table 8). The LDA/NL geometry represents a better fit to experiment than

**Table 8.** Calculated and experimental geometrical parameters of $C_5H_5^-$, $LiC_5H_5$, and ferrocene

| bond length/Å | LDA/NL | exp.[a] | re[b] | LDA/NL | LDA/NL |
|---|---|---|---|---|---|
| | ferrocene | ferrocene | ferrocene | LiCp | $C_5H_5^-$ |
| C–C | 1.432 | 1.440 ± 0.002 | 1.431 ± 0.005 | 1.421 | 1.418 |
| C–H | 1.091 | 1.104 ± 0.006 | 1.122 ± 0.020 | 1.092 | 1.096 |
| Fe–C | 2.048 | 2.064 ± 0.003 | 2.058 ± 0.005 | 2.123 | |
| Fe–H | 2.829 | | 2.814 ± 0.009 | 2.917 | |
| CH-tilt | 0.7° towards | 1.6(4)° [c] | | 2.6° away | |

[a] [43b], based on electron diffraction measurement. [b] [67] estimate for the equilibrium parameters derived from ED $r_g$ structure with correction assuming harmonic vibrations. [c] [43a] and [45]

results obtained by *ab initio* methods or by the more approximate LDA scheme. The Fe–C, and C–C distances are within .01 Å of the observed values, and this deviation is only slightly larger than the experimental uncertainties. The largest deviation between experiment and theory is in the position of the hydrogen atoms; our optimized C–H distances are slightly too short. Experimentally, the hydrogen atoms were found to tilt towards the metal with an angle of 1.6°, [43a, 45] while the calculated tilting angle in 0.7° in the same direction.

There is no experimental data for the geometry of the free Cp ring or for LiCp; the calculated values are also listed in Table 8.

The calculated and experimental geometrical parameters for $Bz_2Cr$, $BzCr(CO)_3$ and $Cr(CO)_6$ are listed in Table 9. The gas-phase data recorded for $Bz_2Cr$ is in very good agreement with our calculated geometrical parameters. The CrH distance is the only parameter that significantly differs from the empirical values. However, the empirical Cr–H distance carries a large experimental uncertainty.

The gas phase structure of $BzCr(CO)_3$ has recently been determined by Kukolich et al. from microwave spectrum [46]. Unfortunately, this microwave measurement provides bond lengths with an uncertainty of 0.01–0.02 Å. Previous electron diffraction measurements by Chiu et al. found six equivalent CC bonds at 400 K, suggesting free internal rotation at that temperature [47]. This qualitative difference from our calculated geometry makes it difficult to compare this experiment with our calculations. Therefore, we included in Table 9 only the solid state data determined by Rees [48]. Since we can only compare the solid state structures of $BzCr(CO)_3$ and $Cr(CO)_6$, with the calculated ones, we cannot comment on the absolute deviation between the calculated and empirical values for the free molecules. Generally, the calculated structure of $BzCr(CO)_3$ agrees well with the empirical data. The difference between the two distinct types of CC bonds is reproduced well. The calculated $CrC_{carbonyl}$ distance of $BzCr(CO)_3$ is too long, compared to the solid state data. However, this bond

**Table 9.** Calculated and experimental geometrical parameters of di-benzene-chromium, chromium-benzene-tricarbonyl, and chromium-hexacarbonyl

| bondlength/Å | LDA/NL | exp. | bondlength/Å | LDA/NL | exp. |
|---|---|---|---|---|---|
| | | Bz$_2$Cr[a] | | | BzCr(CO)$_3$[b] |
| C–C | 1.418 | 1.423 ± 0.002 | CrC$_{ring}$ | 2.222 | 2.223, 2.233, 2.243 |
| C–H | 1.096 | 1.090 ± 0.005 | CrC$_{carbonyl}$ | 1.864 | 1.845 |
| Cr–C | 2.146 | 2.150 ± 0.002 | CO | 1.164 | 1.159, 1.157 |
| Cr–H | 2.958 | 2.935 ± 0.040 | CC | 1.405 | 1.406, 1.407 |
| CH-tilt | 3.2° | 5° towards Cr | | 1.424 | 1.424, 1.422 |
| | | | CH | 1.093 | 1.106, 1.113, 1.109 |
| | | Cr(CO)$_6$[c] | CrCO angle | 179.8° | 177.9°, 178.5° |
| CrC | 1.917 | 1.918 | CCrC$_{carbonyl}$ | 87.2° | 89.14, 86.37 |
| CO | 1.154 | 1.141 | CCC | 120.0° | 120.07°, 119.8°, 120.13° |
| | | | CCH | 120.1° | 119.72° |

[a] Bz$_2$Cr gas phase electron diffraction, uncorrected $r_g$ values. [68] [b] Neutron diffraction solid state data. [48] [c] Neutron diffraction solid state data. [66]

length is substantially longer according to both gas phase experiments: 1.86 Å [46] and 1.863 Å [47], which compare well with the calculated 1.864 Å value. The calculated C–O distances are too large compared to the solid state values for both BzCr(CO)$_3$ and Cr(CO)$_6$, while the CrC distance of Cr(CO)$_6$ is well reproduced. In both Bz$_2$Cr and BzCr(CO)$_3$, the benzene ring maintains the planarity of the carbon framework, and all of the hydrogen atoms are tilted equally towards the metal by 3.2 and 2.6 degrees, respectively. The direction of the hydrogen tilt is in qualitative agreement with empirical observations. All CC bonds in Bz$_2$Cr are of the same length, while short and long CC bonds alternate in BzCr(CO)$_3$.

## 4.2 Vibrational Frequencies and Revised Assignments

The comparison between experimental frequencies and our calculated values is the only direct way in which we can obtain information about the accuracy of the computational method. Also, the unusually large deviations between calculated and observed frequencies for some vibrational modes suggest uncertainties in the empirical assignments. We begin our discussion with ferrocene.

The theoretical and experimental frequencies of ferrocene are given in Table 10. Our calculated frequencies, obtained by the LDA method, show good agreement with the experimental fundamental frequencies. The previously accepted assignments were reported by Bodenheimer and Low [49], who confirmed the main features of the original controversial assignments by Lippincott and Nelson [50]. Our frequency calculations have further substantiated the main features of these frequency assignments. The assignment of the ring vibrations is based on the near coincidence of the infrared and Raman bands.

This coincidence is explained by the weakness of the interaction between the two ring vibrations, as well as by the in-phase and out-of-phase combinations being mutually exclusive in the Raman and in the infrared spectrum.

We have modified the original empirical assignments for the $E_2'$ and $E_2''$ symmetry blocks listed in Table 10, since the CH-wagging and the planar ring distortion frequencies deviate by 100–200 cm$^{-1}$ from the experimental frequencies that were based on the original assignments. We have left out the experimental 1191 and 1189 cm$^{-1}$ frequencies from the fundamentals, and re-assigned the empirical $v_{25}$, $v_{27}$ and $v_{31}$, $v_{33}$ frequencies accordingly. Based on the available theoretical and empirical data, we assign the 1058, and 897 cm$^{-1}$ frequencies to $v_{24}$ and $v_{25}$, calculated to be 1014 and 838 cm$^{-1}$, respectively. Similarly, in the $E_2''$ symmetry block, the experimental frequencies of 1055 and 885 cm$^{-1}$ correspond to $v_{30}$ and $v_{31}$: 1025 and 845 cm$^{-1}$, based on LDA calculations. Margl et al. reported vibrational frequencies and assignments very similar to ours, based on first-principle molecular dynamics calculations [51].

**Table 10.** Calculated harmonic and observed fundamental frequencies (and intensities)[a] of ferrocene.

| | Sym. and no.[b] | LDA | Exp.[c] | Sym. and no. | LDA | Exp.[c] |
|---|---|---|---|---|---|---|
| $A_1'$ | 1 | 3161 | 3110 | $E_1'$    17 | 3155 (8) | 3077 |
|      | 3 | 1086 | 1102 | 18, 20 | 1371 (5) | 1410 |
|      | 2 | 791 | 814 | 20, 18 | 978 (27) | 1005 |
|      | 4 | 305 | 309 | 19 | 808 (7) | 855 |
|      |   |   |   | 21, [22] | 489 (23) | 492 |
| $A_1''$ | 5 | 1209 | 1255 | 22, [21] | 163 (1) | 179 |
|      | 6 | 44 | 44 | | | |
|      |   |   | $E_2'$   23 | 3138 | 3100 |
| $A_2'$ | 7 | 1210 | 1250 | 26, [24] | 1318 | 1356 |
|      |   |   |   | 24, [26] | 1014 | 1058[d] |
| $A_2''$ | 8 | 3162 (15) | 3103 | 25, [27] | 838 | 897[d] |
|      | 10 | 1088 (10) | 1110 | 27, [25] | 790 | [d] |
|      | 9 | 777 (92) | 820 | 28, [25] | 562 | 597 |
|      | 11 | 458 (34) | 478 | | | |
|      |   |   | $E_2''$   29 | 3139 | 3085 |
| $E_1''$ | 12 | 3153 | 3086 | 32 [30, 33] | 1337 | 1351 |
|      | 13, 15 | 1370 | 1410 | 30 [32] | 1025 | 1055[d] |
|      | 15, 13 | 966 | 998 | 31 [33, 34] | 845 | 885[d] |
|      | 14 | 770 | 844 | 33, [31] | 814 | [d] |
|      | 16 | 362 | 389 | 34, [31] | 560 | 569 |

[a] Frequencies in cm$^{-1}$, infrared absortivities in parentheses, in km/mol. [b] Our mode description is based on potential energy distribution over symmetry coordinates. We have indicated significant minor contributions in square brackets. In some cases two coordinates contribute almost equally to the normal modes; accordingly, two numbers are indicated for these modes. [c] [49] The experimental frequencies were obtained from solid state measurements; therefore the lower frequency values are to be regarded as only approximate. [d] Revised assignments, see text. The 1191 ($E_2'$) and 1189 ($E_2''$) cm$^{-1}$ observed frequencies are not assigned to fundamentals.

Considering the two modes whose empirically assigned frequencies were left out of Table 10, a similar $C_5H_5^-$ frequency was accurately predicted by our calculations at 1012 cm$^{-1}$, and observed at 1020 cm$^{-1}$. Also, a frequency shift of 170 cm$^{-1}$, as suggested by the empirical assignments, is not likely for planar vibrations, especially not in the $e_2$ symmetries, which do not involve skeletal vibrations. Further, the revised assignment significantly reduces the deviation of the calculated CH-deformation frequencies, making them comparable with the deviations of served in other symmetries.

All of the frequencies calculated by the LDA method for the non-degenerate modes agree well with the experimental results. Also, the calculated frequencies are systematically smaller than the experimental values. It is particularly remarkable that all of the skeletal vibrational frequencies have been accurately reproduced, with an average deviation of 15 cm$^{-1}$. The deviation of the calculated frequencies from the experimental ones is somewhat larger than that for benzene. This increased deviation can be attributed to a number of factors; one of them being uncertainties in the frequencies determined by experiment. Different studies report frequencies for ferrocene that vary by as much as 40 cm$^{-1}$ for some normal modes [43c]. Further, for benzene, anharmonic corrections for most frequencies were also taken into account for the experimental frequencies. Some uncertainty is associated with the fact that several frequencies are active in solid state only where the site symmetry is reduced. The lower site symmetry is the result of a somewhat distorted geometry, caused by intermolecular interactions: the rings are staggered by about 10°. Accordingly, these normal frequencies are somewhat different from the frequencies of $D_{5h}$ symmetry which retains ferrocene.

We continue with the comparison of the empirical and theoretical frequencies of the benzene complexes. The infrared spectra of $Bz_2Cr$ was first reported by Snyder in 1959 [52]. The Raman spectrum $Bz_2Cr$ was reported by Schäfer et al. [53], and Cyvin and co-workers published a corresponding harmonic force field [54]. This paper presented the latest frequency assignments for $Bz_2Cr$, and is therefore our starting point for the discussion of the frequency assignments shown in Table 11.

For $Bz_2Cr$, the $A_{1g}$, $E_{1g}$, and $E_{2g}$ symmetry vibrations are Raman active, while the $A_{2u}$ and $E_{1u}$ symmetry vibrations are active in the infrared absorption spectrum. These selection rules hold only for the vapour phase and solution spectra, whereas, all *gerade* vibrations are Raman active and all *ungerade* vibrations are infrared active, in the solid state, due to the lower site symmetry in the crystal. The vibrations that are only active in the solid state are expected to be weak in intensity. The assignments of the CH-stretching frequencies is very uncertain, the observed values ranging from 2900 to 3050 cm$^{-1}$, while the highest and lowest calculated frequencies are only 20 cm$^{-1}$ apart. A similar anomaly is present for the frequencies of the CD vibrations of perdeuterated $Bz_2Cr$. These vibrations range from 2122 to 2278 cm$^{-1}$, while the calculated numbers range from 2264 to 2292 cm$^{-1}$. This observation can only be explained by anharmonic effects. We have modified the assignments of twelve frequencies

**Table 11.** Calculated harmonic and observed fundamental frequencies[1] of di-benzene-chromium

| | Bz$_2$-h$_{12}$Cr LDA/geo/NL | exp | Bz$_2$-d$_{12}$Cr LDA/geo/NL | exp |
|---|---|---|---|---|
| $A_{1g}$ | 259.4 | 277[b] | 241.6 | |
| | 762.1 | 791[b] | 579.8 | 566[c] |
| | 958.3 | 970[b] | 912.0 | 920[c] |
| | 3088.7 | 3053[b] | 2291.9 | 2267[c] |
| $A_{2g}$ | 1288.5 | | 1001.9 | 1021[c] |
| $B_{1g}$ | 1106.9 | | 783.8 | |
| | 1389.3 | 1308[c] | 1387.9 | |
| $B_{2g}$ | 601.1 | | 534.8 | |
| | 874.1 | | 698.8 | |
| | 974.4 | | 929.0 | 955[c] |
| | 3069.0 | | 2264.4 | 2125[c] |
| $E_{1g}$ | 329.4 | 335[b] | 303.4 | |
| | 805.9 | 811[b] | 627.2 | |
| | 981.8 | 999[b] | 773.1 | 802[c] |
| | 1397.5 | 1430[b] | 1251.3 | 1282[c] |
| | 3086.5 | | 2282.8 | 2212[c] |
| $A_{2u}$ | 454.6 | 456[a] | 409.4 | 408[a],421[a] |
| | 760.1 | 794[c] | 614.3 | |
| | 957.2 | 971[a] | 912.6 | 929[a] |
| | 3087.7 | 3047[a] | 2290.7 | 2315[a] |
| $B_{2u}$ | 1111.4 | 1115[a], 1142[c] | 786.5 | 826[c]? |
| | 1394.3 | | 1393.8 | |
| $E_{2g}$ | 429.4 | 409[c], 400[b] | 385.1 | |
| | 610.6 | 604[b] | 578.6 | 566[c] |
| | 869.6 | 910[b] | 689.5 | 699[c] |
| | 1110.4 | 1143[b] | 812.6 | 835[c] |
| | 1484.2 | 1508[b] | 1444.4 | |
| | 3075.7 | | 2269.1 | 2212[c] |
| $A_{1u}$ | 45.6 | | 41.5 | |
| | 1287.5 | 1294[a] | 1001.1 | |
| $B_{1u}$ | 595.8 | | 534.1 | |
| | 864.8 | 868[a] 851[a] | 686.8 | |
| | 983.9 | 1014[a] | 936.6 | |
| | 3068.7 | | 2265.8 | 2260[a] |
| $E_{1u}$ | 140.7 | 171[c]? | 128.9 | |
| | 490.0 | 490[c] | 480.1 | 481[a],479[c] |
| | 828.7 | 836[a] | 642.2 | 669[a],664[c] |
| | 985.6 | 999[c] | 780.1 | 802[c] |
| | 1394.9 | 1426[c] | 1248.0 | 1271[c] |
| | 3086.1 | 3047[a] | 2281.7 | 2278[a] |
| $E_{2u}$ | 363.9 | | 338.1 | |
| | 595.1 | | 554.5 | |
| | 805.6 | | 629.9 | |
| | 1095.7 | 1115[a] | 803.5 | |
| | 1438.1 | | 1391.7 | |
| | 3076.2 | | 2267.0 | |

[1] Frequencies in cm$^{-1}$. [a] [52] [b] [53] Solid state data, wavenumbers should be considered as approximate. [c] [54]

of $Bz_2Cr$ and its deuterated analogue. In the following, we discuss the assignments of the $Bz_2Cr$ frequencies, refering to the frequencies of the non-deuterated molecule unless otherwise indicated.

The assignments of the 811 cm$^{-1}$ Raman line in the $E_{1g}$ symmetry species is labelled uncertain in the empirical assignments; this assignment is confirmed by the calculation. In the $E_{2g}$ representation, the observed band at 1508 cm$^{-1}$ was originally not assigned to any fundamental vibration, while the 1631 cm$^{-1}$ band was assigned to it instead. Our calculation could not confirm this assignment or the assignment of the corresponding vibration of the deuterated molecule. Since not all the observed Raman bands are listed in the literature for the deuterated molecule, we could not find any alternative for this molecule. The frequency observed at 152 cm$^{-1}$ is not likely to be the empirical value for the internal ring rotation in $A_{1u}$ symmetry. Our calculations have reproduced a similar frequency for ferrocene very well – (both the observed and calculated values being 44 cm$^{-1}$); therefore we do not believe that the empirical assignment is correct. The similarities between the ring-ring interactions of $Bz_2Cr$ and ferrocene suggests that this frequency should be closer to the ferrocene value. The similarity between the ring interactions of $Bz_2Cr$ and ferrocene is not only confirmed by this study but is also indicated by the observed spectrum. The observed 152 cm$^{-1}$ band might be the second overtone of the internal ring rotation frequency. This is also allowed by the symmetry rules, since $A_{1u} \times A_{1u} \times A_{1u} = A_{1u}$; and the calculated harmonic frequency of the second overtone would be 136.8 cm$^{-1}$. The difference between the theoretical estimate and the observed frequency could be explained by the large anharmonicity of this mode. In the same $A_{1u}$ symmetry no observed frequency had originally been assigned to the in-plane CH-bending vibration, whereas the calculation suggests that the band observed at 1294 cm$^{-1}$ should be selected. The observed 456 and 490 cm$^{-1}$ frequencies are interchanged between $A_{2u}$ and $E_{1u}$, compared to the original assignments. There is an unusually large difference between the observed and calculated frequency for the $b_{1g}$ CC-stretching frequency. This frequency is related to the benzene $b_{2u}$ CC-stretching vibration, which is recognized to be a pathological case for quantum mechanical calculations of vibrational frequencies. This example demonstrates that, despite the generally good agreement between the LDA-calculated and observed frequencies, we have to exercise caution not to assign bands purely on the basis of matching frequencies.

The vibrational frequencies of $BzCr(CO)_3$ have been recorded by several groups. As opposed to the $Bz_2Cr$ spectra that were recorded on older instruments, English, Plowman and Butler recorded the observed spectra with a high-resolution instrument, using solution and solid state samples [55]. These authors were mainly interested in the CO region of the spectrum, therefore not all regions of the spectrum is listed are their publication. In our frequency assignments we use as many of the frequencies reported by English et al. as possible. For frequencies that were not reported by English et al., we refer to studies by Adams et al. [56], Bisby et al. [57] and Schäfer et al. [58].

An important outcome of our frequency assignment is the confirmation of all of the frequency identifications made by English et al. in the CO region [55]. In our assignments, the parallel and perpendicular CrCO-bendings are systematically interchanged with respect to those of English et al.; this might be attributed to differences in notation. In our assignments, a CrCO bend is considered parallel if it is in the symmetry plane of the molecule.

Also, our normal coordinate analysis revealed that the Cr–CO-, ring–Cr–C- and the C–Cr–C-bending coordinates are strongly coupled in E symmetry. In $A_1$ symmetry, the C–Cr–C umbrella opening mode is coupled with the Cr–C–O parallel bend. For details of the assignments the reader is referred to Table 12, in which we also included mode descriptions based on the criterion of potential energy distribution. The frequency related to the $B_{2u}$ C–C stretching vibration is overestimated, as it is for free benzene and for $Bz_2Cr$.

The observed frequencies at 45 or 46 wavenumbers in the solid state deserves special attention. This frequency has previously been neither assigned to any fundamental vibrations nor explained. In line with our calculation, we suggest that this vibration is the overtone of the internal rotation frequency. This explanation is also consistent with the symmetry rules, as the first overtone of $a_2$ is $a_1$. Our experience shows that the calculation usually reproduces these low frequency vibrations fairly accurately.

The geometry, force field, and harmonic frequencies of $Cr(CO)_6$ calculated by density functional methods have been reported previously [8, 25, 58]. The frequencies calculated with our current procedure are listed in Table 13, along with the experimental data. All CO-stretching frequencies are systematically underestimated by about $50 \text{ cm}^{-1}$. This error is clearly related to the too-long CO distance of the reference geometry. If we could improve the reference geometry, the fit of the calculated CO frequencies to experiment would be better. The only other noticeable deviation is that of the $f_{1u}$ Cr–C–O bending frequency, calculated to be $687 \text{ cm}^{-1}$ while observed at $668 \text{ cm}^{-1}$; this error may be related to the anharmonic contribution to the frequency. When these frequencies are not included the average deviation is only $2.46 \text{ cm}^{-1}$, which is in the range of the anharmonicity of these vibrations. Even the overall average deviation of $12 \text{ cm}^{-1}$ is excellent agreement. This exceptionally good agreement between the observed and predicted frequencies is a further indication that the method combining the LDA/NL reference geometry with the LDA force field is the most accurate approach to the calculation of the vibrational frequencies of transition metal systems.

## 4.3 Comparison of Force Fields

In this section, we discuss the differences and similarities between the force fields of the free $Cp^-$ ring, ferrocene and LiCp, as well as between that of benzene, $Bz_2Cr$ and $BzCr(CO)_3$. The corresponding force constants are listed in Table 14 and 15, respectively. We also compare our force constants with the empirically determined force constants, to the extent they are available. Here, we discuss the force constants in the valence coordinate representation, which is more appropriate for chemical interpretations of interactions; they will primarily be used in the present section, where we study the relation between chemical bonding and force constants. We use standard notation in most cases: $R$ stands for CC stretche, $r$ for CH-stretche, $\beta$ for CH in-plane deformations, $\gamma$ for CH out-of-plane deformations. We also use $s$ for CO- and $u$ for metal-carbon stretching coordinates. The numbered internal coordinates are linear combinations of skeletal coordinates, they are explained in the text and listed in the original reference [10c, 10d, 59]. The internal coordinates for the benzene complexes are shown on Figs. 4a–b, and 5a–b, as examples.

**Cp complexes.** In comparing the ring force constants between $Cp_2Fe$, $CpLi$ and $Cp^-$ we start with the CC-stretch. The numerical values for the diagonal CC-stretching force constant of $Cp_2Fe$, $CpLi$ and $Cp^-$ are calculated to be between the force constant of benzene and the $C_2$–$C_3$ constant of butadiene. The CC-bond strength and-force constant increase in the following order: butadiene [60] (5.092 mdyn/Å), ferrocene (5.542 mdyn/Å), $LiC_5H_5$ (5.758 mdyn/Å), $C_5H_5^-$ (5.857 mdyn/Å), and benzene [10c, 10d] (6.619 mdyn/Å). This trend also follows increasing bond order, as well as decreasing bond length. The lengthening of the CC bond in ferrocene, compared to $C_5H_5^-$, is the consequence of orbital interactions between the metal and the rings. Both the back-donation of metal $d$-electrons into the $\pi^*$ orbitals of the Cp-ring and the donation of $\pi$-electrons into the unoccupied $d$-orbitals on the metal decreases the CC bond order and, consequently, the CC bond strength and stretching force constant. Similar interactions are not possible between Li and the ring, since there is no $d$-orbital in the valence shell of Li. Therefore, the CC bond length in LiCp changes by only 0.003 Å, compared to $C_5H_5^-$, while for ferrocene the elongation is .014 Å. Also, the CC stretching constant of $LiC_5H_5$ is closer to that of $C_5^-$ than to that of ferrocene.

Except for those constants that are zero by symmetry for the planar ring, all CC coupling constants in LiCp and $Cp^-$ are similar in magnitude and sign, and differ in many cases from those of $Cp_2Fe$. Thus, the interaction constant in ferrocene between the CC stretch and the in-plane ring deformation ($R_3q_{41}$) is significantly smaller than corresponding constant of the free $Cp^-$ ring, as a consequence of lowered CC-bond strength. The $R_3q_{49}$ constant, that represents the CC interaction with the metal–ring stretch, is almost an order of magnitude larger for ferrocene than for LiCp. The positive sign of this interaction constant indicates that a stretch of the metal-ligand distance strengthens the CC bond as donation and back-donation is reduced. Such a strong effect is not present in LiCp, which is kept together mainly by electrostatic interactions. Ferrocene also exhibits a strong coupling between the CC stretch and the Cp ring tilt through $R_3q_{56}$. This is understandable, since ring-tilt influences the degree of donation and back-donation, and thus the CC stretch.

The CH-bond length, bond strength, and -stretching constant exhibit the opposite trend to that found for the CC bonds. In the series $C_5H_5^-$, $LiC_5H_5$, and

Table 12. The vibrational frequencies[1] of benzene-chromium-tricarbonyl

| | Bz$_2$-h$_6$Cr(CO)$_3$ LDA/geo/NL | exp | Bz$_2$-d$_6$ Cr(CO)$_3$ LDA/geo/NL | exp | Bz$_2$-Cr($^{13}$CO)$_3$ LDA/geo/NL | exp | |
|---|---|---|---|---|---|---|---|
| $A_1$ | 97.5 | 108[a,s] | 97.1 | 108[a,s] | 97.3 | 108[a,s] | umbrella opening |
| | 287.8 | 295[a,s], 302[a] | 275.5 | 291[a] | 286.2 | 294[a,s], 310[a] | skeletal stretch |
| | 476.2 | 478[a] | 474.5 | 477[a] | 467.8 | 470[a] | Cr–C stretch |
| | 655.6 | 653[a] | 670.6 | 671[a] | 640.7 | 641[a] | Cr CO bend para |
| | 751.5 | 790[b], 783[c] | 555.3 | 586[a,s] | 750.5 | | CH wag |
| | 956.5 | 979[b], 977[d] | 911.4 | 930[b] | 956.5 | | CC stretch |
| | 1129.6 | 1150[b] | 799.5 | 815[b] | 1129.6 | | CH rock |
| | 1388.8 | 1316[b?] | 1387.6 | | 1388.8 | | CC stretch |
| | 1972.4 | 1970, 1976[b] | 1972.4 | 1975[a] | 1926.2 | 1931[a] | CO stretch |
| | 3140.5 | 3110[b] | 2330.0 | 2285[b] | 3140.5 | | CH stretch |
| $A_2$ | 22.6 | 46[a,s] ($2\nu_{11}$) | 21.4 | 45[a,s] ($2\nu_{11}$) | 22.5 | 45[a,s] ($2\nu_{11}$) | int. rot., pucker |
| | 396.8 | | 396.3 | | 384.8 | | Cr–CO bend perp |
| | 623.7 | | 547.4 | | 623.5 | | ring puckering |
| | 917.0 | | 743.1 | | 917.0 | | CH wag[e] |
| | 994.3 | | 948.5 | | 994.2 | | ring trig. def. |
| | 1302.4 | | 1012.8 | | 1302.4 | | CH rock |
| | 3117.9 | | 2302.5 | | 3117.9 | | CH stretch |

| E | | | | | | |
|---|---|---|---|---|---|---|
| 86.3 | 61[a,s], 93[c] | 86.1 | 60[a,s], 93[c] | 85.9 | 62[a,s] | skeletal bends |
| 123.5 | 131[a,s] | 119.0 | 130[a] | 123.2 | 130[a,s] | BzCrC, CrCO perp |
| 306.7 | 328[a,s], 335[a] | 289.7 | 313[c] | 305.5 | 329[a,s], 334[a] | ring tilt 1 |
| 402.2 | 423[a] | 362.9 | 381[c] | 401.7 | 425[a,s] | ring asym. torsion |
| 460.3 | 471[a] | 459.6 | 472[a] | 449.4 | 465[a] | Cr–C stretch |
| 525.6 | 532[a], 543[b] | 524.2 | 529[a] | 511.0 | 517[a] | Cr–CO bend para |
| 597.0 | 615[a] | 567.2 | 572[a] | 595.0 | | ring asym. def. |
| 625.9 | 624[a], 637[b] | 617.7 | 622[a] | 614.6 | 608, 617[a] | Cr–CO bend perp |
| 831.5 | 842[d,s], 800[d] | 648.0 | 635[b] | 831.4 | | CH wag |
| 889.8 | 904[b] | 709.1 | | 889.8 | | CH wag |
| 993.4 | 1017[b] | 784.2 | 806[b] | 993.4 | | CC stretch[f] |
| 1127.6 | 1161[b], 1150[d] | 827.7 | 850[b], 829[c] | 1127.5 | | CH rock |
| 1409.7 | 1448[b] | 1259.9 | 1292[a], 1283[c] | 1409.7 | | CH rock[f] |
| 1483.0 | 1519[b] | 1436.3 | 1465[b] | 1483.0 | | CC stretch |
| 1917.9 | 1918[b], 1904[a] | 1917.9 | 1918[b], 1904[a] | 1873.3 | 1860[a] | CO stretch |
| 3124.2 | 3023[b] | 2305.4 | 2215[b] | 3124.2 | | CH stretch |
| 3136.4 | 3090[b] | 2320.8 | 2249[b] | 3136.4 | | CH stretch |

1 Frequencies in cm$^{-1}$. a [55] b [56] c [57] d [11] eStrongly coupled with the ring puckering for the deuterated isotopomer. fStrong mixing between CH rock and CC stretch for the deuterated isotopomer. sTaken from solid state measurements, wavenumbers should be considered only approximate.

**Table 13.** Calculated harmonic and observed fundamental frequencies of chromium-hexacarbonyl.

|  | $Cr(CO)_6$ LDA/geo/NL | exp[a] |
|---|---|---|
| $A_{1g}$ | 2087.9 | 2139.2 |
|  | 384.9 | 379.2 |
| $E$ | 1998.3 | 2045.2 |
|  | 393.5 | 390.6 |
| $F_{1g}$ | 359.6 | 364.1 |
| $F_{1u}$ | 1977.9 | 2043.7 |
|  | 687.4 | 668.1 |
|  | 449.7 | 440.5 |
|  | 94.3 | 97.2 |
| $F_{2g}$ | 534.4 | 532.1 |
|  | 89.1 | 89.7 |
| $F_{2u}$ | 511.5 | 510.9 |
|  | 66.3 | 67.9 |

[a] [69]

**Table 14.** Selected valence force constants of ferrocene, $C_5H_5^-$, $LiC_5H_5$.

|  | ferrocene | $C_5H_5^-$ | $LiC_5H_5$ | description |
|---|---|---|---|---|
| $R^2/2$ | 5.5418 | 5.8567 | 5.7577 | CC stretch |
| $R_1/R_2$ | 0.4877 | 0.4558 | 0.4646 |  |
| $R_1R_6$ | 0.0746 |  |  |  |
| $R_1\beta_1$ | 0.1096 | 0.1084 | 0.1160 |  |
| $R_1\beta_3$ | 0.0111 | 0.0097 | 0.0084 |  |
| $R_3q_{41}$ | 0.3286 | 0.4547 | 0.4459 |  |
| $R_3q_{49}$ | 0.3334 |  | 0.0468 |  |
| $R_3q_{56}$ | − 0.2781 |  | − 0.0957 |  |
| $r_2/2$ | 5.4148 | 5.2415 | 5.3730 | CH stretch |
| $r_1\gamma_1$ | 0.0719 |  | 0.0221 |  |
| $r_1q_{41}$ | − 0.1076 | − 0.1736 | − 0.1390 |  |
| $\beta^2/2$ | 0.3999 | 0.3921 | 0.4035 | CH planar def. |
| $\beta_1\beta_2$ | 0.0084 | 0.0126 | 0.0094 |  |
| $\beta_1\beta_3$ | − 0.0095 | − 0.0107 | − 0.0111 |  |
| $\beta_1q_{42}$ | − 0.0576 | − 0.0670 | − 0.0687 |  |
| $\gamma^2/2$ | 0.4038 | 0.2876 | 0.3431 | CH out-of-plane def. |
| $\gamma_1\gamma_2$ | − 0.0215 | − 0.0445 | − 0.0320 |  |
| $\gamma_1\gamma_3$ | 0.0122 | 0.0309 | 0.0288 |  |
| $\gamma_1q_{44}$ | − 0.2158 | − 0.2063 | − 0.2296 |  |
| $\gamma_1q_{49}$ | 0.0360 |  | − 0.0166 |  |
| $q_{41}^2/2$ | 1.5590 | 1.5820 | 1.6187 | ring planar def. |
| $q_{41}q_{45}$ | 0.0375 |  |  |  |
| $q_{44}^2/2$ | 0.5124 | 0.5835 | 0.5898 | ring out-of-plane def |
| $q_{49}^2/2$ | 3.2430 |  | 1.0838 | Cp–Fe stretch (49–50) |
| $q_{49}q_{50}$ | 0.5213 |  |  |  |
| $q_{51}^2/2$ | 0.3101 |  |  | Cp–Fe–Cp bend (51–52) |
| $q_{51}q_{56}$ | − 0.0447 |  |  |  |
| $q_{53}^2/2$ | 0.0031 |  |  | ring internal rotation |
| $q_{54}^2/2$ | 1.3543 |  | 0.4792 | ring tilt (54–57) |
| $q_{54}q_{55}$ | − 0.0796 |  |  |  |

[a]Coordinates defined in [59]. Units are mdyn/Å, mdyn Å/rad and mdyn/rad for stretches, stretch-bend and bends, respectively.

ferrocene, the CH-bond lengths decrease as 1.096, 1.092, 1.091 Å, respectively, while the corresponding force constants increase to 5.242, 5.372 and 5.425 mdyn/Å, respectively. This trend is also a consequence of the metal-ligand interaction. Thus, as the CC bond order and length decreases, the orbitals describing the adjacent CH bonds gain more $s$ character from carbon, increasing the CH bond strength. The CH stretching force constants in the Cp-systems are comparable to or slightly larger than that in ethylene, calculated to be 5.249 mdyn/Å, by our LDA calculations.

The coupling constants between CH stretches and other coordinates are very small, except for $r_1\gamma_1$ and $r_1q_{41}$. The $r_1\gamma_1$ interaction constant is zero in $Cp^-$ by symmetry, and takes on a positive value in ferrocene and LiCp. The $r_1q_{41}$ coupling constant is the only off-diagonal force constant of significance in $Cp^-$. Its absolute value is reduced somewhat in LiCp and even more in $Cp_2Fe$.

The in-plane ring-deformation constants $(q_{41}^2/2)$, the out-of-plane ring-deformation constants $(q_{44}^2/2)$, the in-plane CH-deformation constants $(\beta^2/2)$ as well as the coupling constants involving $q_{41}$, $q_{44}$ and $\beta$, are not significantly different in the three systems. Also, off-diagonal force constant that vanish by symmetry in $Cp^-$ are insignificant in LiCp and $Cp_2Fe$.

The value of the diagonal $\gamma$ force constant, representing the out-of-plane CH deformation, is almost 50% larger in ferrocene than in the free ring. This is the coordinate for which the corresponding diagonal force constants show the largest percentage difference between the free Cp-ring and ferrocene. A reason for the increased CH-wagging force constant can be found, if one considers the interaction between the hydrogen $1s$ orbital and the $\pi^*$ orbital at the distorted position. In ferrocene, the energy of the $\pi^*$ orbital is substantially increased, compared to the free $Cp^-$ ring. Therefore, this stabilizing interaction is less noticeable in $Cp_2Fe$ than in the free $Cp^-$ ring. The lack of stabilization increases the force constants of ferrocene. The value of the diagonal $\gamma$ force constant for LiCp is also somewhat larger than that of $Cp^-$.

Most coupling constants between ring coordinates on two different rings in ferrocene are small. Significant coupling effects are seen only between the planar ring distortions $q_{41}q_{45}$.

Comparing the skeletal force constants for LiCp and ferrocene, it is apparent that both the metal–Cp stretching and the tilting force constants are about three times as large for ferrocene. This is the consequence of the different electronic structure of the two compounds, the five covalent Fe–C bonds being stronger than the mainly electrostatic $Li^+Cp^-$ interaction. Further, for ferrocene, there is strong coupling between the two FeCp stretches. Also, the tilting coordinates show appreciable interactions with the tilting of the other ring, and with the skeletal bendings. The positive sign of the interaction constant between the two skeletal stretches indicates that the stretching of one metal-ring bond allows increased orbital interaction between the metal and the other ring.

We have also compared our data with the experimentally determined force constants. Since the only reported ferrocene force field is based on treating the $C_5H_5^-$ rings as separate entities, a direct comparison is not possible. The normal coordinate analysis of the entire ferrocene complex was reported by Brunvoll, Cyvin, and Schafer [61]. These authors, however, did not list their refined force constants, but only the result obtained by using the force field of $C_5H_5^-$. Therefore, we compared their CH- and CC-stretching force constants for the free Cp ring with our calculated data. The calculated diagonal constants compare well with the empirical data, especially the CC-stretching constants, that is 5.565 and 5.857 mdyn/Å, based on experiment and our calculations, respectively. The CH-stretching constant is 5.242 mdyn/Å by our calculations and 5.102 mdyn/Å from experiment. The difference is related to the too-short CH-bond length at the calculated reference geometry as well as the effect of anharmonicity. The empirical interaction constants do not compare well with the calculated values. Based on our calculation, the $R_1R_3$ force constant ( $-.020$ mdyn/Å) should be smaller by an order of magnitude than the $R_1R_2$ coupling constant (0.464). In the experimental result, the $R_1R_3$ term is about 20% larger in absolute magnitude, and of opposite sign ( $-0.365$ mdyn/Å), compared to the $R_1R_2$ coupling constant (0.292 mdyn/Å). The CH-stretching interaction constants are not in agreement with our calculated data. The empirical $r_1r_2$ and the $r_1r_3$ interaction constants are $-0.066$ and 0.037 mdyn/Å, while they are 0.018 and 0.004 based on our calculations. Previous experience with force field calculations shows that the signs and magnitudes of the calculated LDA force constants resemble the physical values [10c, 10d]. Therefore, this comparison also shows that it is very difficult to determine reliable force constants empirically for medium size molecules.

Hartley and Ware have determined the metal-Cp stretching constants of ferrocene, empirically from solid state frequencies [62]. These values are in good agreement with our calculated data. The empirically diagonal and interaction force constants are 3.15 and 0.56 mdyn/Å, respectively, while they are 3.243 and 0.521 mdyn/Å according to LDA calculations.

**Benzene complexes**. We continue with the comparison of the empirical and calculated force constants of $BzCr(CO)_3$ and $Bz_2Cr$. We are also interested in comparing the analogous force constants of benzene, $Bz_2Cr$, $BzCr(CO)_3$, and $Cr(CO)_6$. Such a comparison should help to assess the degree to which force constants can be transferred between different transition metal complexes. The question of transferability is of crucial importance for the development of molecular mechanics and dynamics force fields applicable to transition metal complexes. We shall also provide an analysis of the changes upon complexation in the benzene and CO force constants based on qualitative orbital theory. Attention will finally be given to the coupling force constants betwen the ligand coordinates and the skeletal coordinats, as well as between the CO and benzene force constants in $BzCr(CO)_3$.

The internal coordinates were selected so as to make it physically meaningful to compare force constants of different molecules. For the free and the coordinated benzene ring, the internal coordinates selected were based on suggestions by Pulay et al. [10b, 13c]. The C–Cr–C bending coordinates of $Cr(CO)_6$ were chosen in a way that allowed analogous definitions in the case of

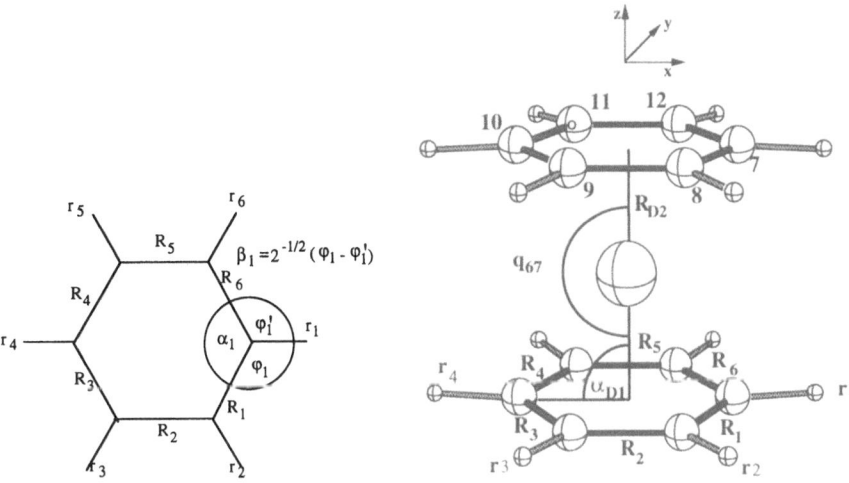

a    Internal coordinates for benzene          b          The internal coordinates of Cr(Bz)$_2$

**Fig. 4a, b.**

BzCr(CO)$_3$. The skeletal internal coordinates, e.g. ligand-metal stretching, ligand tilting, etc., are based on our previous recommendations [59]. The significant force constants of benzene, Bz$_2$Cr, BzCr(CO)$_3$ and Cr(CO)$_6$ are compared in Tables 15–17.

The empirical information about the force constants is very limited for both Bz$_2$Cr and BzCr(CO)$_3$. We have to emphasize that for the direct comparison of force constants one has to use the same internal coordinate systems. (See Section 2.2.) A study by English et al., using accurate frequency assignments provides the most thorough normal coordinate analysis for BzCr(CO)$_3$. This study serves as our reference for the comparison of empirical and calculated potential constants [55]. Prior to this study, the reported CO- and CrC-stretching force constants for BzCr(CO)$_3$ obtained by different authors had ranged from 13.55 to 14.64 mdyn/Å and from 1.6 to 3.88 mdyn/Å, respectively. This wide range of values is the consequence of the different approximations used in the normal-coordinate analysis.

Although, English et al. used compliance constants for the normal coordinate analysis, the compliance field was finally inverted to the more familiar force constant representation. Since the internal coordinate system used here is fundamentally different from that of English et al., the comparison of force constants, although informative, can only be qualitative. The empirical CO-stretching and CO–CO-stretching coupling constants, 15.41 and 0.3, respectively, compare well with the calculated values of 15.09 and 0.23. The difference in the diagonal constants is related to the somewhat too long CO-bond lengths of our reference geometry.

75

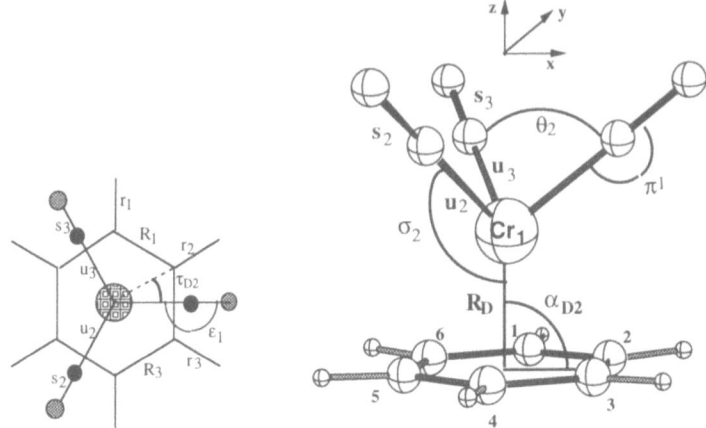

a  Internal coordinates for BzCr(CO)₃     b     Internal coordinates for BzCr(CO)₃

**Fig. 5a, b.**

The calculated CrC-stretching constants are expected to be very accurate since the related frequencies are reproduced well for both BzCr(CO)$_3$ and Cr(CO)$_6$. We calculate the diagonal CrC-stretching force constants to be 2.729 mdyn/Å, while it is 2.4 mdyn/Å according to the experiment. The coupling constants between two CrC stretches are − 0.031 and 0.285 according to theory and experiment, respectively. The differences between the CrC stretching constants cannot be attributed solely to the different internal coordinate systems. We believe, rather, that the differences are related to the lack of information in the empirical determination of force constants; this could only be established by comparing the compliance constants, but those are not reported by English et al.

The bending force constants are in general more seriously effected by the different definitions of internal coordinates, as is the case for the CrCO bending force constants. The perpendicular bending force constants are 0.813 and 0.484 mdyn Å/rad, according to experiment and theory, respectively. Further, the effect of several large coupling constants, neglected in the empirical study, was incorporated into the diagonal constants. For example, the coupling between the C–Cr–C bending and the Cr–C–O bending coordinates was neglected in the empirical study, although some of these constants can be as large in magnitude as half of the diagonal constants themselves. Such important coupling constants are the $\pi_1 q_{46}$ and the $\varepsilon_1 q_{45}$ constants, listed in Table 16.

The empirical data concerning the force constants for Bz$_2$Cr is even more limited. Therefore, we do not make a quantitative comparison with the empirical force constants of Bz$_2$Cr. However, our data clearly support the qualitative observation by Snyder [52] who reported, that CC-stretching force constants of benzene change considerably upon coordination and the CH-out-of-plane constants are close to those of ferrocene.

We begin the comparison of similar force constants in different molecules with the benzene type force constants of free benzene, $Bz_2Cr$, and $BzCr(CO)_3$, listed in Table 15. The most remarkable difference in the stretching force constants is noted for the CC-diagonal and coupling force constants. The diagonal CC-stretching constant of $Bz_2Cr$ is 5.956 mdyn/Å, while the corresponding constant of benzene is 6.952 mdyn/Å. The downward shift in the value of this stretching constant is related to the increased CC-bond length in $Bz_2Cr$, compared to benzene itself. The interaction between the benzene ring and the metal is realized through electron donation from the metal to the $\pi^*$ orbitals of benzene and back-donation from the $\pi$ orbitals of benzene to the unoccupied $d$ orbitals of the metal. Both of these interactions decrease the CC-bond order and -bond strength, explaining why the bond length is increased and the force

**Table 15.** Benzene-like force constants[a]

| pot. term | $Bz_2Cr$ | benzene/LDA | $BzCr(CO)_3$ | $BzCr(CO)_3$ symm. ineqv. |
|---|---|---|---|---|
| $rr/2$ | 5.1994 | 5.210 | 5.3615 | |
| $r_1r_2(o)$ | 0.0024 | 0.006 | 0.0056 | |
| $r_1r_3(m)$ | − 0.0061 | 0.002 | − 0.0029 | |
| $r_1r_4(p)$ | − 0.0016 | 0.003 | − 0.0016 | |
| $r_1R_1$ | 0.0837 | 0.094 | 0.0698 | $r_1R_6 = 0.0695$ |
| $r_1R_2$ | − 0.0108 | − 0.004 | − 0.0174 | $r_1R_5 = -0.0131$ |
| $r_1R_3$ | − 0.0267 | − 0.016 | − 0.0176 | $r_1R_4 = -0.0178$ |
| $r_1\beta_2(o)$ | 0.0052 | 0.009 | 0.0081 | $r_1\beta_6(o) = -0.0077$ |
| $r_1\beta_3(m)$ | − 0.0029 | − 0.008 | − 0.0048 | $r_1\beta_5(m) = 0.0045$ |
| $r_1q_{19}$ | − 0.0769 | − 0.113 | − 0.0737 | |
| $r_1q_{20a}$ | − 0.0917 | − 0.103 | − 0.0728 | |
| $R_1R_1/2$ | 5.9558 | 6.952 | 6.3061 | $R_2R_2/2 = -5.7869$ |
| $R_1R_2(o)$ | 0.5324 | 0.651 | 0.5468 | |
| $R_1R_3(m)$ | − 0.0411 | − 0.381 | − 0.0861 | |
| $R_1R_4(p)$ | 0.0785 | 0.307 | 0.0738 | |
| $R_1\beta_1$ | .1573 | 0.162 | 0.1696 | $R_2\beta_2 = 0.1680$ |
| $R_1\beta_3$ | − 0.0056 | − 0.018 | − 0.0069 | $R_2\beta_6 = -0.0078$ |
| $R_1\beta_4$ | 0.0104 | 0.021 | 0.0137 | $R_2\beta_5 = 0.0133$ |
| $R_1q_{20a}$ | 0.1242 | 0.125 | 0.1081 | $R_2q_{20a} = -0.2590$ |
| | | | | $(R_1q_{20a}^*2)$ |
| $\beta\beta_1/2$ | 0.4723 | 0.491 | 0.4854 | |
| $\beta_1\beta_2(o)$ | − 0.0086 | 0.008 | 0.0065 | |
| $\beta_1\beta_3(m)$ | 0.0086 | − 0.011 | − 0.0100 | |
| $\beta_1\beta_4(p)$ | − 0.0031 | − 0.002 | − 0.0022 | |
| $\beta_2q_{20a}$ | − 0.0578 | − 0.067 | − 0.0669 | |
| $q_{19}q_{19}/2$ | 1.2111 | 1.213 | 1.2489 | |
| $q_{20a}q_{20a}/2$ | 1.2486 | 1.207 | 1.2342 | |
| $\gamma_1\gamma_2/2$ | 0.4217 | 0.442 | 0.4408 | |
| $\gamma_1\gamma_2(o)$ | − 0.0340 | − 0.072 | − 0.0513 | $\gamma_1\gamma_6(o) = -0.0347$ |
| $\gamma_1\gamma_3(m)$ | 0.0069 | 0.006 | 0.0048 | |
| $\gamma_1\gamma_4(p)$ | − 0.0113 | − 0.022 | − 0.0130 | |
| $\gamma_1q_{28}$ | − 0.1357 | − 0.158 | − 0.1392 | |
| $\gamma_1q_{29a}$ | − 0.1782 | − 0.146 | − 0.1712 | |

[a]Coordinates defined in [59]. Units are mdyn/Å, mdyn Å/rad and mdyn/rad.

constant is decreased. A similar effect plays a role in the change of the CC-bond length and stretching force constants of $BzCr(CO)_3$. However, in case of $BzCr(CO)_3$, the picture is complicated by the presence of the three CO groups opposite to the ring. All CC bonds of the ring *trans* to the CO groups are shorter and stronger, while the others are longer and weaker. We note that the presence of the CO group significantly reduces the positive charge on the Cr atom. While the calculated charge of Cr in $Bz_2Cr$ is 1.7, it is only 0.95 in $BzCr(CO)_3$.

We also observed a significant decrease in the magnitude of the CC-stretching coupling constants. The *ortho* and *meta* coupling constants of free benzne are large, as is characteristic of aromatic and conjugated systems. The significantly decreased coupling in the complexes is a result of the strong donation into the ring $\pi^*$ orbitals.

The remaining force constants for the planar vibrations of benzene are not significantly changed after complexation. However, the out-of-plane ring-deformation force constants change significantly. While the $q_{28}$ diagonal force constant is decreased, the $q_{29}$ force constant is increased. The $q_{28}$ internal coordinate describes the motion that brings the planar ring into a chair conformation through alternating expansion and contraction of the carbon-chromium distance.

During the study of ferrocene, we noticed a significant increase in the CH out-of-plane displacement force constant, compared to the free ring. This increase was explained by the lack of a stabilising interaction between the $\pi^*$ orbitals and the $1s$ orbital of the hydrogen in ferrocene, due to the higher energy of the $\pi^*$ orbital in the complex. Since the $\pi^*$ orbitals are of higher energy in the free benzene ring than in the free $Cp^-$ ring, relative to the CH-bonding orbitals, similar stabilizing interactions are not present either in the free benzene ring or in the complex. Therefore, we have not noted any change in the values of the CH-wagging force constants of benzene upon complexation.

We continue our discussion with a comparison of the force constants of $BzCr(CO)_3$, and $Cr(CO)_6$, listed in Table 16. The CO diagonal stretching force constant is smaller for $BzCr(CO)_3$, than for $Cr(CO)_6$. At the same time, an opposite trend is apparent for the CrC-stretching constants. Both trends can be explained by stronger back-donation to the CO ligands in $BzCr(CO)_3$, where only three strong $\pi$-acceptors compete for electrons. The diagonal Cr–C–O-bending constants are increased somewhat in $BzCr(CO)_3$, compared to $Cr(CO)_6$. Even more pronounced differences are noted in the C–Cr–C bending constants. These trend can also be related to the stronger back-donation in $BzCr(CO)_3$.

The diagonal and off-diagonal skeletal force constants are listed in Table 17. The value of the ligand-metal stretching force constant of $Bz_2Cr$ is very close to that of ferrocene; they are 3.119 and 3.243 mdyn/Å, respectively. The corresponding force constant for $BzCr(CO)_3$ is significantly smaller, 2.298 mdyn/Å. In both benzene complexes there is strong coupling between the CC stretch and the benzene–chromium stretch, and the coupling constant is positive. Upon the increase of the metal–ligand distance, the force on the CC bond acts to reduce

**Table 16.** Force constants for the CrCO internal coordinates in $BzCr(CO)_3$ and $Cr(CO)_6$[a]

| pot. term | $BzCr(CO)_3$ | $Cr(CO)_6$ | pot. term | $BzCr(CO)_3$ | $Cr(CO)_6$ |
|---|---|---|---|---|---|
| $s_1 s_1/2$ | 15.0934 | 16.2166 | $\pi_1 \pi_1/2$ | 0.4740 | 0.4345 |
| $s_1 s_2$ | 0.2264 | 0.1801 | $\pi_1 \pi_2$ | 0.0044 | 0.0049 |
| $s_1 u_1$ | 0.7579 | 0.6566 | $\pi_1 \varepsilon_2$ | − 0.0031 | 0.0056 |
| $s_1 u_2$ | − 0.0606 | − 0.0603 | $\pi_1 q_{43}$ | 0.1552 | 0.1706 |
| $s_1 \pi_1$ | 0.0205 | 0.000 | $\pi_1 q_{44}$ | − 0.0627 | − 0.0754 |
| $s_1 \pi_2$ | 0.0035 | − 0.0026 | $\pi_1 q_{46}$ | − 0.1338 | |
| $s_1 q_{43}$ | 0.0220 | 0.0044 | $\pi_1 q_{50}$ | − 0.0257 | |
| $s_1 q_{44}$ | − 0.0387 | − 0.0031 | $\varepsilon_1 \varepsilon_1/2$ | 0.4838 | 0.4345 |
| $s_1 q_{48}$ | − 0.1299 | | $\varepsilon_1 \varepsilon_2$ | − 0.0182 | − 0.0049 |
| $s_1 q_{50}$ | 0.0533 | | $\varepsilon_1 q_{45}$ | − 0.2286 | 0.1828 |
| $u_1 u_1/2$ | 2.7290 | 2.2361 | $\varepsilon_1 q_{47}$ | 0.0902 | |
| $u_1 u_2$ | − 0.0305 | − 0.0128 | $\varepsilon_1 q_{49}$ | 0.0040 | |
| $u_1 \pi_1$ | − 0.0515 | 0.0000 | $q_{43} q_{43}/2$ | 0.7289 | 0.9467 |
| $u_1 \pi_2$ | 0.0310 | 0.0194 | $q_{43} q_{48}$ | − 0.0770 | |
| $u_1 \varepsilon_2$ | − 0.0213 | 0.0194 | $q_{44} q_{44}/2$ | 1.0724 | 0.8059 |
| $u_1 q_{44}$ | 0.0368 | 0.0005 | $q_{44} q_{46}$ | − 0.3644 | |
| $u_1 q_{46}$ | − 0.0264 | | $q_{44} q_{50}$ | 0.0195 | |
| $u_1 q_{48}$ | 0.2106 | | $q_{46} q_{46}/2$ | 0.9315 | |
| $u_1 q_{50}$ | − 0.1484 | | $q_{46} q_{50}$ | 0.0514 | |

[a]Coordinates defined in [59]. Units are mdyn/Å, mdyn Å/rad and mdyn/rad for stretches, stretch-bend and bends, respectively.

**Table 17.** Selected skeletal force constants[a]

| | $Bz_2Cr$ | $BzCr(CO)_3$[b] | | $Bz_2Cr$ | $BzCr(CO)_3$ |
|---|---|---|---|---|---|
| $R_1 q_{61}$ | 0.2711 | 0.2112 | $q_{28} q_{69}$ | | − 0.0096 |
| $R_1 q_{62}$ | 0.0028 | | $q_{29a} q_{64}$ | | − 0.0760 |
| $R_2 q_{64}$ | 0.2176 | 0.1034 | $q_{61} q_{61}/2$ | 3.1186 | 2.2976 |
| $R_2 q_{66}$ | 0.0127 | | $q_{61} q_{62}$ | 0.2816 | |
| $R_2 q_{68}$ | 0.0244 | | $q_{63} q_{63}/2$ | 1.2250 | 1.0365 |
| $q_{19} q_{69}$ | | 0.0157 | $q_{63} q_{65}$ | − 0.1524 | |
| $q_{20a} q_{64}$ | | 0.0561 | $q_{63} q_{67}$ | − 0.0409 | |
| $\gamma_1 q_{61}$ | 0.0680 | 0.0405 | $q_{67} q_{67}/2$ | 0.2797 | |
| $\gamma_1 q_{63}$ | − 0.0832 | − 0.0819 | $q_{69} q_{69}/2$ | 0.0041 | 0.0011[c] |

[a]Coordinates defined in [59]. Units are mdyn/Å, mdyn Å/rad and mdyn/rad for stretches, stretch-bend and bends, respectively. [b]The internal coordinates are numbered according to the definition of $Bz_2Cr$. [c]The definition of internal rotation is different for $Bz_2Cr$ and $BzCr(CO)_3$.

the CC bond length, as a result of decreased electron donation from the metal to the $\pi^*$ orbitals of the benzene ring.

The values of the force constants of $Bz_2Cr$ for ligand–metal–ligand bending-, ligand tilting-, and internal rotation are very close to those of ferrocene. The corresponding constants of $BzCr(CO)_3$ are somewhat different, a feature that might also be related to (unavoidable) differences in the definitions of the internal coordinates.

The interaction force constants between the benzene ring and the $Cr(CO)_3$ fragment of the $BzCr(CO)_3$ complex are listed in Table 18. In general, we note that there are many important coupling constants between these two fragments of the complex. The most important are the interaction constants between the CC-stretching constants and the CrC- and CO-stretching constants. The $R_1$ CC stretch is strongly coupled with both the CrC ($u_2$) and the CO ($s_2$) bond stretching coordinates in the positions *trans* to the CC bond. The $R_2$ CC stretch is coupled with CrC ($u_1$) and CO ($s_1$) bond stretches, which are in a quasi-eclipsed position to the $R_2$ CC bond. The couplings between the CC-stretching coordinate and the Bz–Cr–C bendings are also significant; for example, the $R_1 q_{46}$ and $R_1 q_{47}$ coupling constants. Also, the longer CC bonds ($R_2$, $R_4$, $R_6$) are coupled with the C–Cr–C bending coordinates. This interaction is shown in the values of the $R_2 q_{43}$ and $R_2 q_{44}$ coupling constants. The planar and out-of-plane of the benzene deformations ring are also coupled with the Bz–Cr–C- and C–Cr–C- deformation coordinates.

As the values in Table 18 indicate, the coupling between the benzene ring and the rest of the complex cannot simply be neglected, which was often the practice in the course of the normal-coordinate analysis of this compound, based on observed vibrational frequencies. As mentioned before, the normal-coordinate analysis from the reverse vibrational problem cannot provide all of the significant coupling constants of the harmonic force field; this is simply due to the paucity of experimental information. Quantum mechanical calculations are necessary to obtain information about the significant coupling constants.

### 4.3.1 Transfer of force constants and kinematic coupling

The vibrational frequencies of the free Cp ring and corresponding frequencies of ferrocene are remarkably different for some modes. Similar differences are seen between the frequencies of benzene and of the complexed benzenes. Differences in the frequencies can arise for two reasons. The most obvious reason is the differences in the force fields, but the same force field can produce different frequencies if the reduced masses are different, due to the different kinetic energy matrix. Brunovoll and co-workers realized that the reduced masses of benzene and that of the complexed benzene are significantly different, and this – according to Brunvoll and co-workers – is the main reason why the frequencies differ from those of benzene [63].

During the first normal-coordinate studies of ferrocene, the authors treated the Cp rings as separate entities; [50] therefore, the effect of the changing reduced mass was not introduced in this treatment. The revision of the normal-coordinate analysis of ferrocene by Brunvoll and co-workers, based on the treatment of the ferrocene complex as a whole, has demonstrated that similar kinematic coupling effects play a significant role in the normal vibrational frequencies for ferrocene as well [43c]. This study also concluded that the differences between the frequencies of ferrocene and those of the free Cp- ring are

**Table 18.** Coupling constants between the Bz ring and CO in $BzCr(CO)_6$[a]

| pot. term | $BzCr(CO)_3$ | pot. term | $BzCr(CO)_3$ | pot. term | $BzCr(CO)_3$ |
|---|---|---|---|---|---|
| $r_1s_1$ | 0.0031 | $R_1q_{47}$ | 0.0800 | $q_{20}q_{44}$ | − 0.0290 |
| $r_1s_2$ | − 0.0034 | $R_2s_1$ | 0.0553 | $q_{20}q_{46}$ | − 0.0352 |
| $r_1s_3$ | 0.0011 | $R_2s_2$ | − 0.0119 | $\gamma_1s_1$ | 0.0023 |
| $r_1u_1$ | 0.0017 | $R_2u_1$ | − 0.0393 | $\gamma_1s_2$ | 0.0025 |
| $r_1u_2$ | − 0.0011 | $R_2u_2$ | 0.0136 | $\gamma_1s_3$ | − 0.0222 |
| $r_1u_3$ | 0.0068 | $R_2u_3$ | 0.0136 | $\gamma_1u_1$ | − 0.0052 |
| $r_1q_{43}$ | − 0.0039 | $R_2\pi_1$ | − 0.0147 | $\gamma_1u_2$ | − 0.0065 |
| $r_1q_{44}$ | 0.0023 | $R_2\pi_2$ | − 0.0134 | $\gamma_1u_3$ | 0.0115 |
| $r_1q_{45}$ | 0.0050 | $R_2\varepsilon_2$ | 0.0063 | $\gamma_1\pi_2$ | 0.0084 |
| $r_1q_{46}$ | − 0.0031 | $R_2\pi_3$ | − 0.0134 | $\gamma_1\varepsilon_3$ | 0.0026 |
| $r_1q_{47}$ | − 0.0041 | $R_2\varepsilon_3$ | − 0.0063 | $\gamma_1q_{43}$ | 0.0085 |
| $R_1s_1$ | − 0.0065 | $R_2q_{43}$ | − 0.0295 | $\gamma_1q_{44}$ | − 0.0018 |
| $R_1s_2$ | 0.0394 | $R_2q_{44}$ | 0.0754 | $\gamma_1q_{45}$ | 0.0112 |
| $R_1s_3$ | − 0.0065 | $R_2q_{46}$ | 0.0378 | $\gamma_1q_{46}$ | − 0.0043 |
| $R_1u_1$ | 0.0138 | $R_2q_{48}$ | 0.1938 | $\gamma_1q_{47}$ | − 0.0130 |
| $R_1u_2$ | − 0.0368 | $\beta_1s_3$ | − 0.0047 | $q_{28}\varepsilon_1$ | − 0.0016 |
| $R_1u_3$ | 0.0138 | $\beta_1u_3$ | 0.0022 | $q_{28}\varepsilon_2$ | − 0.0045 |
| $R_1\pi_1$ | 0.0119 | $\beta_1\pi_3$ | 0.0028 | $q_{29a}s_1$ | 0.0125 |
| $R_1\varepsilon_1$ | 0.0028 | $\beta_1q_{45}$ | 0.0022 | $q_{29a}s_2$ | − 0.0062 |
| $R_1\pi_2$ | − 0.0159 | $\beta_1q_{46}$ | − 0.0055 | $q_{29a}s_3$ | − 0.0062 |
| $R_1\pi_3$ | 0.0119 | $\beta_1q_{48}$ | 0.0035 | $q_{29a}u_1$ | − 0.0215 |
| $R_1\varepsilon_3$ | − 0.0030 | $q_{19}\varepsilon_1$ | 0.0147 | $q_{29a}u_2$ | 0.0107 |
| $R_1q_{43}$ | 0.0097 | $q_{20}s_1$ | − 0.0019 | $q_{29a}u_3$ | 0.0107 |
| $R_1q_{44}$ | 0.0010 | $q_{20a}s_2$ | 0.0010 | $q_{29a}\pi_1$ | 0.0082 |
| $R_1q_{45}$ | 0.0016 | $q_{20}u_1$ | − 0.0031 | $q_{29a}q_{44}$ | − 0.0317 |
| $R_1q_{46}$ | − 0.0461 | $q_{20}u_2$ | 0.0016 | $q_{29a}q_{46}$ | 0.0216 |

[a] Coordinates defined in [59]. Units are mdyn/Å, mdyn Å/rad and mdyn/rad.

mainly attributed to kinematic coupling effects, whereas the ring force constants are essentially identical for the two compounds.

In order to test the effect of kinematic coupling, we calculated the harmonic frequencies of ferrocene, based on the valence force constants of the free $Cp^-$ ring but without coupling constants between the skeletal and ring coordinates. This calculation has confirmed that some frequencies change due to the mixing of skeletal and internal ring vibrations as a result of the kinematic effects. However, the change in the frequencies is significantly less in magnitude than in previous studies. The $E_1''$ CH wagging frequency shifts from 620 cm$^{-1}$ to 669 ($E_1''$) and 659 ($E_1'$) cm$^{-1}$, whereas Brunvoll et al. report a shift from 625 to 673 ($E_1''$) and 859 ($E_1'$) cm$^{-1}$. According to our results, the $A_2''$ CH wagging frequency shifts from 629 cm$^{-1}$ to 641 ($A_1'$) and 643 ($A_2''$) cm$^{-1}$, as opposed to the shift from 710 cm$^{-1}$ to 941 ($A_1'$) and 892 ($A_2''$) cm$^{-1}$ reported previously.

The differences between the results of this and previous studies underline the importance of a proper representation of the skeletal modes. On the basis of our normal-coordinate analysis with the complete force field, the contribution of the skeletal coordinates in those frequencies that exhibit shifts is not more than 5% in terms of the potential energy distribution. This contribution is far too small to explain several frequency shifts of more than $200 \, \text{cm}^{-1}$. Also, the quantum mechanical data showed a 50% increase in the CH wagging force constant, underlining the fact that the shifts of the CH wagging frequencies are mainly due to the increased force constant rather than of skeletal-ring mode interactions.

Our normal-coordinate analysis of $Bz_2Cr$ also reveals that there is no significant kinematic coupling between the skeletal and ring coordinates, and the frequency shifts of the benzene vibrations are related to changes in the force constants, and not to the kinematic coupling effect suggested by Brunvoll et al. [61]. Coupling between coordinates depends on the values of both the coupling force constants and the off-diagonal elements of the kinetic energy matrix. In our representation, the off-diagonal G-matrix elements between the ring-metal-stretch and the CC stretch are zero, while it can be as much as 20–30% of the diagonal elements, if metal–carbon bonds are used to represent the skeletal vibrations. This large coupling is related to the interdependence of the CC and metal–carbon bonds. It is likely, therefore, that the strong coupling between skeletal and ring vibrations reported by Brunvoll et al. are related to the artefacts of their internal coordinate representation.

# 5 Conclusions

This set of studies establishes a systematic computational procedure for calculating the force constants and vibrational frequencies of transition metal complexes. We showed that accurate reference geometries are crucial for obtaining good results. The LDA/NL method provides geometries very close to experimental ones. These geometries can be used in the force field and frequency calculations, for which the LDA method is already quite accurate. Our study also points out that special attention should be given to the internal coordinate representation of the potential constants, when the force constants of complexed and free molecules are compared. We suggest physically meaningfull internal coordinates to represent the skeletal distortions.

We have reproduced the vibrational frequencies of benzene, transition metal carbonyls and transition metal complexes of benzene and cyclopentadiene fairly accurately. Our calculations indicate some ambiguities in the original empirical frequency assignments for ferrocene and dibenzene-chromium, for which we suggest alternatives. Our calculations confirm the frequency assignments for $BzCr(CO)_3$.

The comparison of the force constants of the free aromatic rings with those of the complexes show substantial changes in the force fields upon complexation. Major changes were seen in the force constants for out-of-plane distortions and CC-stretching coordinates.

This set of calculations represents a successful new area of application for density functional theory. These calculations provide the first complete force fields for ferrocene, dibenzene-chromium, and benzene-chromium-tricarbonyl.

*Acknowledgments*: This investigation was supported by the Natural Sciences and Engineering Research Council of Canada (NSERC) as well as the Donors of the Petroleum Research Fund, administered by the American Chemical Society (ACS-PRF #27023-AC23). The Academic Computing Service of the University of Calgary is acknowledged for access to the IBM-6000/RISC facilities.

# 6 References

1. Versluis L, Ziegler T (1988) J Chem Phys 88: 322
2. Fournier R, Andzelm J, Salahub, DR (1989) J Chem Phys 90: 6371
3. (a) Handy NC, Tozer DJ, Murray CW, Laming GJ, Amos RD (1993) Isr J Chem 33: 331
   (b) Dunlap I, Andzelm J (1992) Phys Rev A45: 81
   (c) Johnson BG, Frisch MJ (1994) J Chem Phys 100: 7429
4. Fournier R, Papai I (1996) in Chong DP (ed) "Recent Advances in Density Functional Methods", Part I, World Scientific
5. Ziegler T (1991) Chem Rev 91: 651
6. (a) Ziegler T, Rauk A (1977) Theoret Chim Acta (Berl.) 43: 261
   (b) Becke AD ACS Symp Ser (1989) : 394
   (c) Becke AD (1982) J Chem Phys 76: 6037
7. (a) Fan L, Versluis L, Ziegler T, Baerends EJ, Ravenek W (1988) Int J Quantum Chem S22: 173
   (b) Papai I, St-Amant A, Fournier R, Salahub DR (1989) Int J Quantum Chem S23
8. (a) Fan L, PhD Thesis The University of Calgary (1992)
   (b) Fan L, Ziegler T (1992) J Chem Phys 96: 9005
   (c) Fan L, Ziegler T (1992) J Phys Chem 96: 6937
9. (a) Andzelm J, Wimmer E (1992) J Chem Phys 96: 1280
   (b) Sosa C, Andzelm J, Elkin BE, Wimmer E, Dobbs KD, Dixon DA, (1992) J Phys Chem 96: 6630
10. (a) Goodman L, Ozkabak AG, Thakur SN (1991) J Phys Chem 95: 9044
    (b) Pulay P, Fogarasi G, Boggs JE (1981) J Chem Phys 74: 3999
    (c) Bérces A, Ziegler T (1993) J Chem Phys 98: 4793
    (d) Bérces A, Ziegler T (1993) J Chem Phys Lett 203: 592
    (e) Handy NC, Maslen PE, Amos RD, Andrews JS, Murray CW, Laming GJ (1992) Chem Phys Lett 197: 506
11. Schäfer L, Begun GM, Cyvin SJ (1972) Spectochim Acta 28A: 803
12. Schwendeman RH (1966) J Chem Phys 44: 556 ibid (1966) 44: 2115
13. (a) Blom CE, Altona C (1976) Mol Phys 31: 1377
    (b) Pulay P, Fogarasi G, Pang F, Boggs JE (1979) J Am Chem Soc 101: 2550
    (c) Fogarasi G, Pulay P (1985) in: Durig JR (ed) Vibrational Spectra and Structure Elsevier, New York, Vol 14, p 125–219
    (d) Fogarasi G, Zhou X, Taylor PW, Pulay P (1992) J Am Chem Soc 114: 8191
14. Allen WD, Csaszar A (1993) J Chem Phys 98: 2983
15. Bérces A. Ziegler T(1995) J Phys Chem (1995) 99 : 11417
16. (a) El'yashevich MA (1940) Dokl Akad Nauk SSSR 28: 605 (in Russian)
    (b) Wilson Jr EB (1941) J Chem Phys 9: 76
    (c) Wilson Jr EB, Decius JC, Cross PC Molecular Vibrations McGraw-Hill New York, (1955)
17. Baerends EJ, Ellis DE, Ros P (1973) Chem Phys 2: 41

18. Ravenek W in Algorithms and Applications on Vector and Parallel Computers; te Riele HJJ; Dekker Th J; van de Vorst HA (Eds.); Elsevier, Amsterdam, (1987)
19. (a) Boerrigter PM, te Velde G, Baerends EJ (1988) Int J Quantum Chem 33: 87
    (b) te Velde G, Baerends EJ (1992) J Comp Phys 99: 84
20. Krijn J, Baerends EJ (1984) "Fit functions in the HFS-method", Internal Report (in Dutch) Free University of Amsterdam, The Netherlands
21. (a) Snijders GJ, Baerends EJ, Vernooijs P (1982) At Nucl Data Tables 26: 483
    (b) Vernooijs P, Snijders GJ, Baerends EJ (1981) "Slater Type Basis Functions for the whole Periodic System"; Internal report, Free University of Amsterdam, The Netherlands
22. Vosko SH, Wilk L, Nusair M (1980) Can J Phys 58: 1200
23. Becke AD (1988) Phys Rev A 38: 2398
24. Perdew JP (1986) Phys Rev B33: 8822; ibid (1986) B34: 7046
25. (a) Fan L, Ziegler T (1991) J Chem Phys 94: 6057
    (b) Fan L, Ziegler T (1991) J Chem Phys 95: 7401
26. (a) Császár P, Pulay P (1984) J Mol Struct 114: 31
    (b) Pulay P (1980) Chem Phys Lett 73: 393
    (c) Pulay P (1982) J Comput Chem 3: 556
27. Szalay PG, based on the program GEOMO by Pulay. Eotvos University, Budapest, Hungary
28. Bérces A, Ziegler T, Fan L (1994) J Phys Chem 98: 1584
29. Fan L, Versluis L, Ziegler T, Baerends EJ, Ravenek W (1988) Int J Quantum Chem S22: 173
30. Maurer F, Wieser H The University of Calgary, Calgary, Canada
31. Schachtschneider, Shell Development Company Emeryville California, (1960)
32. (a) Hochstrasser RM, Wessel JE, Sung HN (1974) J Chem Phys 60: 317
    (b) Hochstrasser RM, Sung HN, Wessel JE (1973) J Am Chem Soc 95: 8179
    (c) Berman JM, Goodman L (1987) J Chem Phys 87: 1479
    (d) Wunsch L, Metz F, Neusser HJ, Schlag EW (1977) J Chem Phys 66: 386
33. (a) Guo H, Karpulus M (1988) J Chem Phys 89: 4235
    (b) Ozkabak AG, Goodman L, Wiberg KB (1990) J Chem Phys 92: 4115
    (c) Zhou X, Fogarasi G, Pulay P unpublished, (1992)
    (d) Pulay P (1986) J Chem Phys 85: 1703
34. Pongor G, Pulay P, Fogarasi G, Boggs JE (1984) J Am Chem Soc 106: 2765
35. Sellers H, Pulay P, Boggs JE (1985) J Am Chem Soc 107: 6487
36. (a) Ozkabak AG, Goodman L, Thakur SN, Krogh-Jespersen K (1985) J Chem Phys 83: 6047
    (b) Ozkabak AG, Goodman L (1987) J Chem Phys 87: 2564
37. (a) Buijse MA, Baerends EJ (1991) J Chem Phys 93: 4190
    (b) Buijse MA, Baerends EJ (1991) Theor Chim Acta 79: 389
38. Tschinke V, Ziegler T (1991) Theor Chim Acta 81: 65
39. (a) Akiyama M (1980) J Mol Spectrosc 84: 49; (1982) 93: 154
    (b) Dang-Nhu M, Pliva J (1989) Mol Spectrosc 138: 423
40. Goodman L, Ozkabak AG, Wiberg KB (1989) J Chem Phys 91: 2069
41. Hollenstein H, Piccirillo S, Quack M, Snels M (1990) Mol Phys 71: 759
42. Whiffen DH (1955) Philos Trans R Soc London Ser A 248: 131
43. (a) Haaland A (1979) Acc Chem Res 12: 415
    (b) Haaland A, Nilsson JE (1968) Acta Chem Scandinavica 22: 2653.
    (c) Sado A, West R, Fritz HP, Schafer L (1966) Spectrochim Acta 22: 509
44. Doman TN, Landis CR, Bosnich B (1992) J Am Chem Soc 114: 7264
45. Takusagawa F, Koetzle TF (1979) Acta Crystallogr B35: 1074
46. Kukolich SG, Sickafoose SM, Flores LD, Breckenridge SMJ (1994) Chem Phys 100: 6125
47. Chiu N-S, Schäfer L, Seip R (1975) Organomet Chem 101: 331
48. Rees B, Coppens P (1973) Acta Cryst B29: 2515
49. Bodenheimer JS, Low W (1973) Spectrochimica Acta 29A: 1733
50. Lippincott ER, Nelson RD (1958) Spectrochim. Acta 10: 307
51. Margl P, Schwarz K, Blöchl PE (1994) J Chem Phys 100: 8617
52. Snyder RG (1959) Spectrochim Acta 10: 807
53. Schäfer L, Southern JE, Cyvin SJ (1971) Spectrochim Acta 27A: 1083
54. Cyvin SJ, Brunvoll J, Schäfer L (1971) J Chem Phys 54: 1517
55. English AM, Plowman KR, Butler IS (1982) Inorg Chem 21: 338–347
56. Adams DM, Christopher RE, Stevens DC Inorg Chem (1975) 14: 1562
57. Bisby EM, Davidson G, Duce DA (1978) J Mol Structure 48: 93
58. Delley B, Wrinn M, Luthi HP (1994) J Chem Phys 100: 5785

59. Berces A, Ziegler T (1994) J Phys Chem 98: 13233
60. Pulay P, Fogarasi G, Ponger G, Boggs JE, Vargha A (1983) J Am Chem Soc 105: 7037
61. Brunvoll J, Cyvin SJ, Schafer L (1971) J Organometal Chem 27: 107
62. Hartley D, Ware J (1969) J Chem Soc (A) 138
63. Brunvoll J, Cyvin SJ, Schafer L J. Organometal Chem (1971) 27: 69
64. Stoicheff BP (1954) Can J Phys 32: 339
65. Jones LH, McDowell RS, Goldblatt M (1969) Inorg Chem 8: 2349
66. Jost A, Rees B, Yelon WB (1975) Acta Crystallogr B31: 2649
67. Bohn RK, Haaland A, (1966) J Organometal Chem 5: 470–476
68. Haaland A (1965) Acta Chem Scandinavica 19: 41
69. Hedberg L, Lijima T, Hedberg K (1979) J Chem Phys 70: 3224

# Structure and Spectroscopy of Small Atomic Clusters

R. O. Jones

Institut für Festkörperforschung, Forschungszentrum Jülich, D-52425 Jülich, Germany

## Table of Contents

Topics in Current Chemistry, Vol. 182
© Springer-Verlag Berlin Heidelberg 1996

The geometrical arrangement of the atoms is one of the most important properties of any material. In principle, it can be found from calculations of the total energy of the system, but accurate solutions of the Schrödinger equation can be found for relatively few systems, and the density functional formalism – combined with simulated annealing at finite temperatures – provides a method for both calculating energies and avoiding unfavourable minima in the energy surface. We outline some of the methods that can provide experimental spectroscopic information about the structures of atomic clusters, and we show how density functional calculations can aid, in particular, the analysis of photodetachment measurements and provide interesting and unexpected results.

# 1 Introduction

Atomic clusters have been the central topic of so many conferences [1], summer schools [2], and books [3] that it is becoming a cliché to note their importance. "Clusters" mean different things to different people – in particular, the *number* of atoms comprising a "cluster" is a matter of taste or convention – but to many researchers they provide a fascinating area between the physics of atoms and small molecules, on the one hand, and bulk phases on the other. In previous articles, I have surveyed particular aspects of clusters, including their generation and spectroscopy [4] and some of their bonding trends [5]. In the present work, I shall focus on the spatial arrangement of the atoms and how theory can aid the interpretation of experimental data, particularly photoelectron detachment data from negative ions.

The geometrical arrangement of the constituent atoms is one of the most important properties of any material. This may be self-evident to chemists or molecular physicists, as the study of molecules and their interactions implies a knowledge of the atomic positions. It was certainly so for molecular biologists such as Francis Crick, who wrote: "If you want to study function, study structure" [6]. It may be less obvious to those physicists who seek universal rules that apply to *all* systems. In emphasizing the geometrical structure and related properties, we address at the outset problems that are specific to *individual* systems, although structural trends can be quite fascinating.

Clusters are generally considered to be aggregates of one or two elements, so that methods for determining structural information about molecules are obvious candidates for the study of clusters. A long established and one of the most accurate methods – X-ray diffraction – can sometimes be used, and we shall give examples below. Structural information can often be found from the rotations and vibrations of molecules, and infrared and Raman spectroscopy have also been used for this purpose. The improvement in mass separation techniques in recent years has provided new possibilities for studying clusters and their ions. Cluster cations with many thousands of atoms can be detected, and high-resolution spectroscopy can be performed on cluster anions. The analysis of these data often requires a thorough study of the energy surfaces for the cluster and its ions, and one of the goals of the present paper is to discuss the role that density functional calculations can play.

In principle, the stable geometric arrangements of atoms in any material – in the present case, neutral and charged molecules and clusters – can be found from calculations of the total energy $E$ of the system of electrons and ions for a set of nuclear coordinates $\{R_I\}$. If the calculation is performed for *all* possible configurations, the most stable structure is that with the lowest energy. There are two distinct problems associated with this procedure: the calculation of $E$ for *one* geometry, and the determination of the most stable of the possible structures.

It is natural to seek to determine the energy $E$ from the exact wave function $\Psi$ of the system, as we could then calculate not only the total energy

$$E_{GS} = \langle \Psi | \hat{\mathscr{H}} | \Psi \rangle / \langle \Psi | \Psi \rangle, \tag{1}$$

where $\hat{\mathscr{H}}$ is the Hamiltonian describing the interactions in the system, but also many other properties of interest. In practice, however, the numerical effort required to calculate accurate energies increases rapidly as the number of electrons increases. Nevertheless, we shall see that such "ab initio" methods have made valuable contributions to the cluster studies.

The second problem – the determination of the most stable structures amongst the many possible – is at least as difficult, since the number of geometrical configurations grows rapidly as the number of atoms $N$ increases. Alternative schemes for finding low-energy structures are required in systems where the ground state is unknown or there are many local minima, and Kirkpatrick et al. [7] suggested "simulated annealing" at an elevated temperature as a possibility. The analogy to annealing techniques of experimentalists is immediate, since all methods for improving the perfection of a crystal rely on raising the temperature. The kinetic energy of the ions means that the system is less likely to become trapped in the high-lying minima of the energy surface, and slow cooling can result in energetically favourable structures.

The density functional (DF) formalism [8], with a local spin density (LSD) approximation for the exchange-correlation energy, provides a tractable method for performing energy calculations with predictive value in a range of systems. Car and Parrinello [9] showed, moreover, that it could be combined with molecular dynamics (MD) – particularly with the simulated annealing strategy – to give a parameter-free method for calculating electronic properties that makes no assumptions about ground state geometries. The use of finite temperatures allows an efficient sampling of the potential energy surface, and the method has been invaluable in extending our knowledge of the structures of molecules and clusters, as well as liquids and amorphous materials. In the present chapter, I focus on the clusters of some main group elements (groups 13, 15 and 16) and on the analysis of photoelectron spectra of these systems.

R. O. Jones

# 2 Determination of Cluster Geometries

## 2.1 X-ray Diffraction

The group 16 elements provide some of the best characterized atomic clusters. The elements S and Se, in particular, are unique in that many allotropes are molecular crystals comprising regular arrays of well-separated rings of two-fold coordinated atoms. X-ray structure analyses have been performed for $S_n$, $n = 6\text{--}8, 10\text{--}13, 18, 20$ and $Se_n$, $n = 6, 8$ [10, 11], and we show the structures of $S_n$ found in Fig. 1. There are five crystalline modifications of selenium: four comprise $Se_6$ and $Se_8$ rings, and trigonal Se consists of parallel helical chains [10]. Mixed crystals of the form $Se_nS_m$ [12] and a range of sulphur oxides ($S_nO$, $n = 5\text{--}10; S_7O_2, S_{12}O_2$) [13] and ions are also known. The preparation of these clusters has been reviewed by Steudel [11].

The presence of many molecules with well established structures provides an ideal test of any method of calculation, and this is one of the reasons that our

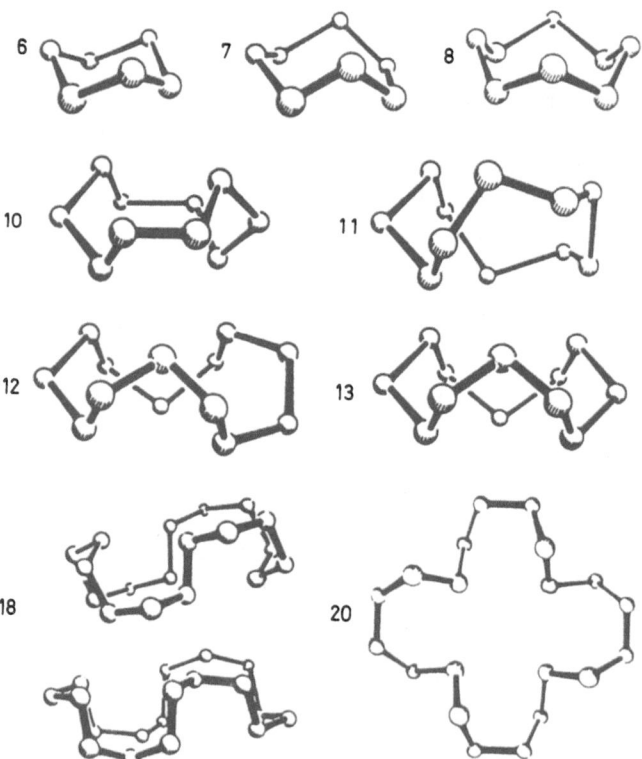

**Fig. 1.** Structures determined by X-ray diffraction for sulphur clusters $S_n$, $n = 6\text{--}8, 10\text{--}13, 18, 20$ (after Steudel [72])

90

first tests of the MD/DF method focused on group 16 clusters and molecules. Unfortunately, the possibility of preparing single crystals is largely restricted to the chain structures that arise in this main group. Although there are elements of other groups with a range of structural forms (boron and phosphorus are familiar examples), there are few for which the structure of clusters can be found from X-ray diffraction. It is then necessary to use other (perhaps less direct) spectroscopic methods.

## 2.2 Vibrational Spectroscopy

Ultraviolet, infrared and Raman spectroscopy have been used widely to study the modes of vibration – and therefore the structural properties – of molecules. These methods have provided invaluable information for many small clusters in the gas phase, particularly diatomic molecules [14]. There are also very precise data on some larger elemental clusters with symmetrical structures (an example is tetrahedral $P_4$ [15]), and recent Raman spectroscopy measurements on $Si_n$ clusters ($n = 4, 6, 7$) [16] have confirmed the predictions of HF-based calculations [17]. Nevertheless, the application to larger clusters in the gas phase has not been widespread to date. One reason has certainly been the difficulty in performing reliable calculations for comparison purposes.

The high symmetry of some of the sulphur structures (Fig. 1) and the presence of a number of twofold degenerate vibrations mean that the infrared and Raman spectra are quite different, so that it is often possible to identify these species, even in mixtures with other molecules of this type in the gas and liquid phases [11]. The correlations found between the vibration frequencies and structural properties (e.g., the relationship in a ring structure between the length of a bond and the dihedral angle) have been very useful in making structural predictions in cases where X-ray diffraction data are presently unavailable.

## 2.3 Photoelectron Spectroscopy

The most stable structure of a particular cluster is often closely related to the structures of clusters of a similar number of atoms, and it is not surprising that the vibrational properties also show relatively smooth changes with changing cluster size. The difficulty of identifying individual clusters by their vibrations alone can be avoided if the clusters can be separated according to their masses, and recent advances in mass spectroscopic techniques have been very important for the study of cluster structures. Mass separation is, of course, usually performed on charged systems.

Examples of work on positively charged clusters are provided by studies of sulphur [18] and phosphorus [19], where clusters could be identified up to 56 and over 6000 atoms, respectively. The trends in the mass abundances may give some indication of the structures to be expected. Measurements of the ionization energies of clusters (for the transition $X_n \to X_n^+$), as in the case of P and As [20], can also provide useful information. Spectroscopy of neutral clusters obtained by photodetachment of mass-selected negative ions is also possible if the neutral clusters are subsequently photoionized. The transient ion signal provides information about the motion of the atoms, particularly if the ionization processes result in a significant change in the geometry. This technique has been applied to small Ag clusters [21]. Photoelectron spectroscopy of negative ions is also a promising method for determining structural information about clusters and is the main focus of the work described here.

The past 10–15 years have seen significant improvements in the negative ion spectroscopy of molecules [22]. In Fig. 2 we give an example of the data that can be obtained and how it can be analyzed. A beam of negatively charged ions is produced ($Al_2^-$ in the present example) and the excess electron is detached, for example by a laser pulse of known energy. If the kinetic energy of electron can be determined, so can its binding energy (BE). The problem for the theorist is illustrated on the right hand side of Fig. 1. The low-lying states of $Al_2^-$ must be found, and the excitation energy to different states of $Al_2$ calculated. If the ionization process is very rapid, the possible final states will have the same geometry as that of the anion. Transitions between states for which the most stable geometries are very similar will evidently lead to sharper peaks in the measured spectrum than transitions between states where the optimum structures are very different.

In addition to these energy differences (vertical detachment energies, VDE), it is sometimes possible to resolve vibrational structure in the binding energies (an example can be seen in Fig. 1 for BE $\sim$ 3.2 eV). The measured vibration frequency can then be compared with calculated values corresponding to the

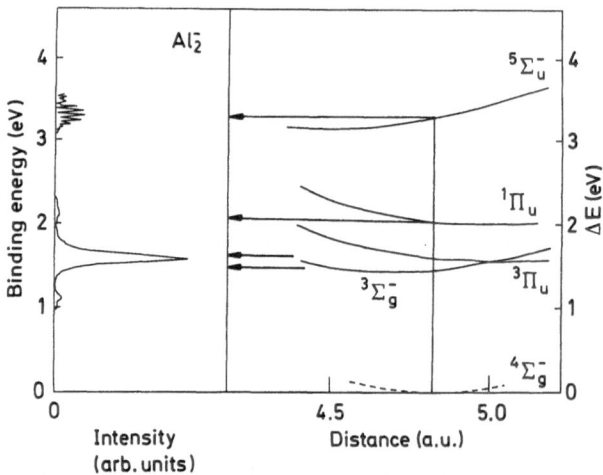

**Fig. 2.** Comparison of measured photoelectron spectrum for $Al_2^-$ [63] with calculated energy differences for transitions between the ground state of $Al_2^-$ and states of the neutral dimer [44]

final state of the transition (here the $^5\Sigma_u^-$ state of $Al_2$). In principle, it should be possible to study the vibrational structure of low-lying states of both the anion and the neutral cluster, although this will depend on the resolution of the measurement. Fig. 1 shows the "interface" between theory and experiment in a diatomic molecule. In larger clusters, of course, the representation of the interatomic coordinates and the comparison between theory and experiment is much more complicated.

There have been numerous photodetachment studies of small cluster anions, and we now give some examples. Noble metal clusters ($Cu_n^-$, $Ag_n^-$, $Au_n^-$, $n = 1$–10) have been studied by Ho et al. [23], who resolved vibrations in all three dimers. Studies of alkali metal cluster anions have included those of $Na_n^-$ ($n = 2$–5), $K_n^-$ ($n = 2$–19), $Rb_{2-3}^-$, and $Cs_{2-3}^-$ [24, 25]. Carbon cluster anions $C_n^-$ have photoelectron spectra that are consistent with linear chains for $n = 2$–9 and monocyclic rings for $n = 10$–29 [26]. Photoelectron spectra of $Sb_n^-$ and $Bi_n^-$ to $n = 4$ [27] show rich vibrational structure for the dimers, and the spectra of the larger clusters could be interpreted in terms of ab initio calculations. The threshold photodetachment (zero electron kinetic energy, ZEKE) spectrum of $Si_4^-$ [28] shows a progression of well-resolved transitions between the ground state of the rhombic anion ($D_{2h}$, $^3B_{2g}$) and vibrational levels of the first excited state of the neutral cluster ($D_{2h}$, $^2B_{3u}$). The measurements were consistent with ab initio predictions of Rohlfing and Raghavachari [29]. As noted by the authors of the experimental work [28], "the role of ab initio calculations in interpreting these spectra cannot be overemphasized. In the absence of experimental force constants for $Si_4$, ab initio calculations are needed to perform any reasonable assignment of the observed vibrational progressions".

The interpretation of the measured spectra is a challenge to the theoretician. This challenge has certainly been met in the case of clusters containing s-valence electrons (alkali metals, alkaline earths, and noble metals), and a detailed review has been given by Bonačić-Koutecký et al. [30]. These authors show, for example, that ab initio calculations for the ground states of $Na_{2-5}^-$ and $Na_{2-5}$ and excited states of the anions reproduce in a quantitative fashion the measured excitation energies and allow an assignment of the anion geometries. We shall show below that MD/DF calculations can provide similar information for main group elements with more complicated valence structures.

# 3 Structures from Total Energy Calculations

The energy of a system is given in terms of its exact wave function $\Psi$ by Eq. (1). If we seek instead a reliable estimate of the wave function, it is common to rely on the Rayleigh-Ritz principle:

$$E' = \langle \Phi | \hat{\mathscr{H}} | \Phi \rangle / \langle \Phi | \Phi \rangle \geq E_{GS}, \tag{2}$$

for an approximate solution $\Phi$, with the equality applying to the exact solution $\Psi$. In practice, this means that improvements in the wave function are reflected in the energy expression (2), with small decreases in $E'$ implying a wave function that is approaching convergence. We now show how this variational principle is applied to molecules and clusters.

## 3.1 Hartree-Fock and Related Methods

One of the earliest approximations for $\Psi$ is due to Hartree, who considered "independent" electrons moving in the field of the other electrons in the system. The variational wave function has the form of a product of single-particle functions, i.e.,

$$\Psi(r_1, r_2, \ldots) = \psi_1(r_1) \ldots \psi_n(r_n) \tag{3}$$

The variational principle then requires that each of the functions $\psi_i(r_i)$ satisfies a one-electron Schrödinger equation of the form

$$\left(-\frac{\hbar^2}{2m} \nabla^2 + V_{\text{ext}} + \Phi_i\right) \psi_i(r) = \varepsilon_i \psi_i(r), \tag{4}$$

where $V_{\text{ext}}$ is the potential due to the nuclei, and the Coulomb potential $\Phi_i$ is given by Poisson's equation and arises from the average field of the other electrons. The state of the system is then defined by the single particle functions $\psi_i(r)$, the eigenvalues $\varepsilon_i$, and the occupancy of the "orbitals" so defined. The Hartree-Fock (HF) approximation, obtained by replacing the product by a single (Slater) determinantal function ("configuration"), leads to an additional non-local "exchange" term $V_x^{\text{HF}}$ in the Schrödinger equation:

$$V_x^{\text{HF}} \psi(r) \equiv \int dr' V_x^{\text{HF}}(r, r') \psi(r') , \tag{5}$$

but the same single-particle picture. HF energies are more accurate than those of Hartree calculations, and the approximation has been an indispensable benchmark in molecular physics since its inception.

In spite of the familiarity of Hartree-Fock calculations, it has long been known that the resulting total energies are inadequate for many purposes. An improved energy results if we take a linear combination of configurations, and this procedure for improving the many-particle wave function – "configuration interaction" (CI) – leads, in principle, to the exact wave function. We have noted that this allows the calculation of many properties of interest, but the numerical effort required increases dramatically with increasing electron number. Quantum Monte Carlo (QMC) methods offer an alternative and very reliable approach to the determination of the wave function of the interacting system. With appropriate choices for the ionic pseudopotential and trial wave function, they have been applied recently to calculations for single geometries of $Si_n$ clusters up to $n = 20$ [31].

## 3.2 Density Functional Formalism

I shall focus here on the method that is the topic of this volume, the density functional (DF) formalism, which is also free of adjustable parameters and can lead to reliable predictions of structures and energy differences in a range of systems. The basic theorems are [32]:
  (1) Ground state (GS) properties of a system of electrons and ions in an external field $V_{ext}$ can be determined from the electron density $n(r)$ *alone*.
  (2) The total energy $E$ is such a functional of the density, and $E[n]$ satisfies the variational principle $E[n] \geq E_{GS}$. The density for which the equality holds is the ground state density, $n_{GS}$.
The usual implementation of this scheme results from the observation of Kohn and Sham [33] that the minimization of $E[n]$ is simplified if we write (we adopt atomic units with $e = \hbar = m = 1$):

$$E[n] = T_0[n] + \int dr\, n(r)(V_{ext}(r) + \tfrac{1}{2}\varphi(r)) + E_{xc}[n] , \qquad (6)$$

where $T_0$ is the kinetic energy that a system with density $n$ would have in the absence of electron-electron interactions, $\varphi(r)$ is the Coulomb potential, and $E_{xc}$ defines the exchange-correlation energy. The choice of kinetic energy term allows us to reduce the numerical problem to the solution of single-particle equations of Hartree-type, with an effective potential related to the functional derivative of $E_{xc}$. The most widely used approximation for $E_{xc}$ is the local spin density (LSD) approximation

$$E_{xc}^{LSD} = \int dr\, n(r)\, \varepsilon_{xc}[n_\uparrow(r), n_\downarrow(r)] , \qquad (7)$$

where $\varepsilon_{xc}[n_\uparrow, n_\downarrow]$ is the exchange and correlation energy per particle of a homogeneous, spin-polarized electron gas with spin-up and spin-down densities $n_\uparrow$ and $n_\downarrow$, respectively.

## 3.3 MD/DF Calculations

If we use the DF formalism (with the LSD approximation for the exchange-correlation energy) to describe the energy surfaces of the system in question, then the determination of the most stable structures must address *two* minimization problems. (i) The DF variational principle requires that – for each geometry – the density be varied to minimize the energy. (ii) The ions must be moved to minimize the energy. We may do both *simultaneously* by viewing $E$ as a function of two interdependent sets of degrees of freedom: the single particle orbitals $\{\psi_i\}$ that lead to the density, and the ionic coordinates $\{R_I\}$ [9],

$$E[\{\psi_i\}, \{R_I\}] = \sum_i \left\langle \psi_i(r) \left| -\frac{\nabla^2}{2} \right| \psi_i(r) \right\rangle + \int dr\, n(r)(V_{ext}(r) + \tfrac{1}{2}\Phi(r))$$

$$+ E_{xc}[n(r)] + \frac{1}{2}\sum_{I \neq J} \frac{Z_I Z_J}{|R_I - R_J|}. \qquad (8)$$

To minimize this function, we follow the trajectories of $\{\psi_i\}$ and $\{R_I\}$ given by the Lagrangian

$$\mathscr{L} = \sum_i \mu_i \int_\Omega dr |\dot{\psi}_i^* \dot{\psi}_i| + \sum_I \frac{1}{2} M_I \dot{R}_I^2 - E[\{\psi_i\}, \{R_I\}]$$

$$+ \sum_{ij} \Lambda_{ij} \left( \int_\Omega dr \psi_i \psi_j^* - \delta_{ij} \right) \tag{9}$$

and the corresponding equations of motion

$$\mu \ddot{\psi}_i(r, t) = -\frac{\delta E}{\delta \psi^*(r, t)} + \sum_k \Lambda_{ik} \psi_k(r, t),$$

$$M_I \ddot{R}_I = -\nabla_{R_I} E. \tag{10}$$

Here $M_I$ are the ionic masses, $\mu_i$ are fictitious "masses" associated with the electronic degrees of freedom, dots denote time derivatives, and the Lagrangian multipliers $\Lambda_{ij}$ are introduced to satisfy the orthonormality constraints on the $\psi_i(r, t)$. From these orbitals and the resultant density $n(r, t) = \sum_i |\psi_i(r, t)|^2$ we evaluate the total energy $E$, which acts as the classical potential energy in the Lagrangian (9). With an appropriate choice of $\mu_i$, the (artificial) Newton's dynamics for the electronic degrees of freedom prevent transfer of energy from the classical to the quantum degrees of freedom over long simulation periods. The method can be applied to both traditional MD applications and simulated annealing.

### 3.4 Computational Details

While DF calculations have been performed with a great variety of numerical techniques and with numerous basis sets to represent the single-particle functions $\psi_i$, most MD/DF calculations to date have used a plane wave (PW) basis expansion and a pseudopotential representation of the electron-ion interaction $V_{ext}$:

$$V_{ext}(r) = \sum_I v_{ps}(r - R_I),$$

$$v_{ps}(r) = \sum_{l=0}^{\infty} v_l(r) \hat{P}_l, \tag{11}$$

where $\hat{P}_l$ is the angular momentum projection operator. Our experience with the prescriptions for determining $v_l(r)$ given by Bachelet et al. [34] and Stumpf et al. [35] indicates that accurate calculations require components of $v_l$ up to at least $l = 2$. The eigenfunctions of the effective Kohn-Sham Hamiltonian are expanded in the PW basis

$$\psi_i(r) = \psi_{jk}(r) = \sum_{n=1}^{M} c_{jG_n}^k \exp(i(k + G_n) \cdot r), \tag{12}$$

leading to the self-consistent eigenvalue problem

$$\sum_n H_{mn} c_{jG_n} = \varepsilon_j c_{jG_m} \,,$$

$$H_{mn} = \delta_{mn} |G_m|^2 + V_{ext}(G_m - G_n) + \Phi(G_m - G_n) + V_{xc}(G_m - G_n) \,, \quad (13)$$

where the index $k$ has been deleted.

We generally use a face-centred-cubic unit cell [lattice constant 15.9 Å] with constant volume $[1000 \text{ Å}^3]$ and periodic boundary conditions (PBC). This supercell geometry leads to a weak interaction between the individual clusters, and the accurate reproduction of the symmetries in $S_6$, $S_8$ and $S_{12}$ [36], for example, shows that the results are insensitive to the choice of boundary conditions. The cut-off energy for the plane wave expansion (11) of the electronic eigenfunctions $\psi_i$ (10.6–14.0 Ry) leads to $\sim 4000–6000$ plane waves for a single point ($k = 0$) in the Brillouin zone (BZ).

The MD parts of the procedure are initiated by displacing the atoms randomly from an arbitrary geometry, with velocities $\dot{\psi}_i$ and $\dot{R}_I$ set to zero, and using a self-consistent iterative diagonalization technique to determine those $\psi_i$ that minimize $E$. With the electrons initially in their ground state, the dynamics (9) generate Born-Oppenheimer (BO) trajectories over several thousand time steps without the need for additional diagonalization/self-consistency cycles. In typical applications, the "mass" $\mu_i$ of the electronic degrees of freedom was 300–1800 a.u., and the MD time step $\Delta t = 1.7–3.4 \times 10^{-16}$ s. If the mean classical kinetic energy of the atoms defines a "temperature" $T$, the energy surfaces are probed by varying $T$, and the minima in the potential energy surfaces are found by reducing $T$ slowly to zero.

# 4 MD/DF Calculations – Structures

To demonstrate the usefulness of the MD/DF approach, we now discuss applications to structure determination in clusters of elements of groups 13, 15, and 16. Clusters of the last two are typically covalently bonded systems. The bulk systems are generally semiconductors or insulators, and there is a substantial energy gap between the highest occupied and lowest unoccupied molecular orbitals. The first, typified by aluminium, show aspects of "metallic" behaviour. One of the advantages of the DF method is that it can be applied with comparable ease to elements of all atomic numbers.

## 4.1 Group 13 Elements: Al, Ga

Work on $Al_n$ has included magnetic properties, [37] ionization thresholds and reactivities, [38], static polarizabilities [39], and measurements of collision-induced dissociation of $Al_n^+$ [40]. The dimer is the best studied of the aluminium clusters, although the nature of the ground state has only recently been established. The best candidates are the $^3\Pi_u$ $(\sigma_g\pi_u)$ and $^3\Sigma_g^-$ $(\pi_u^2)$ states, and experimental work [41] supports theoretical predictions [42, 43] that the $^3\Pi_u$ state is slightly (less than 0.025 eV) more stable. Experimental and theoretical spectroscopic parameters for low-lying states of $Al_2$ are compared in Table 1. The MD/DF calculations [44] agree well with available data for $Al_2$ [14, 41], although the $^3\Pi_u$ state is a little (0.08 eV) less stable than the $^3\Sigma_g^-$ state. The equilibrium separations $r_e$ and vibration frequencies $\omega_e$ agree very well with experiment for both states. The overestimate of the well depth (2.03 eV compared with the experimental value 1.5 eV [14]) is similar to those found in other $sp$-bonded systems [8].

The results of the MD/DF calculations for the larger clusters show the following patterns:

1. The existence of numerous isomers illustrates the relative ease of electron transfer between $\pi$-orbitals (which dominate in the bonding in planar structures) and $\sigma$-orbitals. Planar structures are the most stable for $n < 5$, three-dimensional structures for $n > 5$. There is a transition at $n = 6$ to ground states with minimum spin degeneracy, so that it is essential to incorporate spin in the calculations of lighter clusters. The structural variety is consistent with the "metallic" nature of the elements: The valence $sp$-shells in the atoms are less than half-filled, and the separation in energy between the highest occupied and lowest unoccupied orbitals is usually small.

**Table 1.** Molecular parameters $r_e$ [atomic units], $\omega_e$ [cm$^{-1}$] for low-lying states of $Al_2$, with energies ($\Delta E$) relative to the ground state

| State | | $r_e$ | $\omega_e$ | $\Delta E$ |
|---|---|---|---|---|
| $^3\Pi_u(\sigma_g\pi_u)$: | | | | |
| | (a) | 5.135 | 284.97 | |
| | (b) | 5.150 | 277 | |
| | (c) | 5.19 | 284 | |
| | (d) | 5.095 | 290 | |
| Expt. | (e) | 5.10 | 284.2 | |
| $^3\Sigma_g^-$ $(\pi_u^2)$: | | | | |
| | (a) | 4.687 | 355.15 | + 0.06 |
| | (b) | 4.711 | 343 | + 0.02 |
| | (c) | 4.78 | 340 | + 0.02 |
| | (d) | 4.672 | 340 | − 0.08 |
| Expt. | (e, f) | 4.660 | 350.01 | > 0 |
| $^5\Sigma_u^-$ $(\sigma_g^2\sigma_u\pi_u^2\sigma_g)$: | | | | |
| | (d) | 4.444 | 435 | + 1.59 |
| | (g) | | 450 ± 40 | + 1.6 |

References: (a) [42]: coupled-cluster doubles + ST(CCD). (b) [43]: complete active space SCF/second order CI CASSCF/SOCI. (c) [71]: multireference configuration interaction (MRD-CI). (d) [44]: MD/DF. (e) [41]. (f) [14]. (g) [63].

2. The stable forms in both $Al_n$ and $Ga_n$ clusters are found by capping smaller clusters, as we show in Fig. 3 for $Al_5$ to $Al_{10}$. The structures comprise triangles with patterns of dihedral angles similar to those found in bulk aluminium and in $\alpha$-gallium [44].

3. The bonds in $Ga_2$ are 3–7% *shorter* than those in $Al_2$. It is very unusual for the lighter of two elements in the same main group to have longer bonds, but this general feature of these clusters is also reflected in the structures of the bulk elements. The nearest neighbour separation in f.c.c. Al is 5.411 a.u., and a weighted average over the seven near neighbours in $\alpha$-Ga (5.107 a.u.) is 5.3% less. The bond lengths are consistent with the extent of the valence orbitals [44], the *s*-orbital, in particular, being considerably more compact in Ga than in Al. This is a consequence of "*d*-block contraction", which is particularly pronounced in Ga [44].

### 4.2 Group 15 Elements: P, As

There have been many studies of the crystalline structures of elemental phosphorus [10, 45], and the microscopic structures of the amorphous modifications (red, black, grey vitreous) are still the subjects of considerable attention. Gas phase clusters have been of interest for many years, and Martin [19] has detected mass spectroscopically $P_n^+$ clusters up to $n > 6000$. Nevertheless, little experimental information was available until recently on the structure of clusters with $n > 4$, and this is also true for arsenic clusters, $As_n$.

We have performed MD/DF calculations for neutral and charged clusters of phosphorus and arsenic with up to eleven atoms [46, 47]. The geometries and

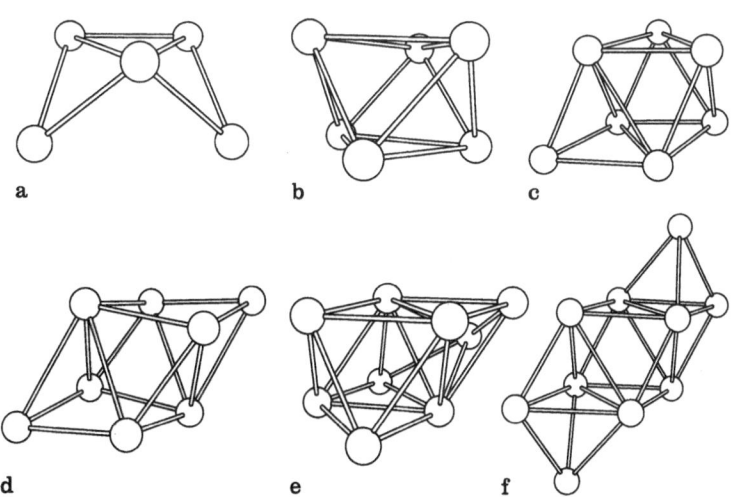

**Fig. 3a–f.** Structures of the most stable isomers calculated for aluminium clusters $Al_n$, $n = 5$–10 [44]

vibration frequencies in $P_2$ and $P_4$ agree well with experimental data, so we may expect reliable predictions for structures that have not yet been established experimentally. The tetrahedral structure is favoured in $P_4$, but there is a large "basin of attraction" for a $D_{2d}$ "roof" structure, i.e., this is the closest minimum for a large region of configuration space. It is also a prominent feature in the low-lying isomers of $P_5$ to $P_8$ [46].

One of the most unexpected results obtained was for $P_8$, where the much-studied cubic ($O_h$) form corresponds to a shallow local minimum in the energy surface. Simulated annealing led, however, to the $C_{2v}$ structure (Fig. 4b), which is much (ca. 1.7 eV) more stable. This "wedge" structure, which may be viewed as a (distorted) cube with one bond rotated through 90°, is a structural unit in violet (monoclinic, Hittorf) phosphorus [48]. A second isomer of $P_8$ ($D_{2h}$, Fig. 4a) is also much more stable than the cubic form. There is a striking analogy between the structures of the $P_8$-isomers and those of the valence isoelectronic hydrocarbons $(CH)_8$. The cubic form of the latter (cubane) has been prepared by Eaton and Cole [49], and can be converted catalytically to the wedge-shaped form (cuneane) [50].

The structures of $P_8$ and the prediction of a $C_{2v}$ isomer as the most stable in $P_6$ have been confirmed by subsequent calculations using HF-based methods [51, 52]. One interesting question that could not be answered definitively by calculations using the LSD approximation is the relative stability of $P_8$ and two $P_4$ tetrahedra, as binding energies can sometimes be overestimated substantially by the LSD approximation [8]. This effect is already apparent in the calculated atomization energies in clusters up to $P_4$, where experimental results are available. The LSD calculations indicated that the $C_{2v}$ isomer of $P_8$ was slightly more stable than two tetrahedra, while the opposite result was found in calculations using HF-based methods [51, 52]. To examine these issues in more detail, Ballone and Jones [47] performed calculations on phosphorus and arsenic clusters with up to 11 atoms using a non-local (gradient corrected) extension of the LSD approximation [53]:

$$E_{xc} = E_{xc}^{LSD} - b \sum_{\sigma} \int dr\, n_{\sigma}^{4/3} \frac{x_{\sigma}^2}{1 + 6bx_{\sigma} \sinh^{-1} x_{\sigma}}. \tag{14}$$

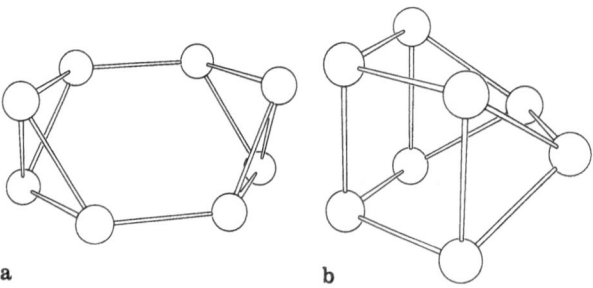

a                                 b

**Fig. 4a, b.** Calculated structures for $P_8$. **a** $D_{2h}$, **b** $C_{2v}$. The latter (cuneane) form is the most stable [46]

where

$$x_\sigma = \frac{|\nabla n_\sigma|}{n_\sigma^{4/3}}; \quad b = 0.0042 \text{ a.u.} \tag{15}$$

The parameter $b$ was adjusted by Becke [53] to reproduce the exchange energies of some closed-shell atoms, and it has been found that this modification to the LSD approximation gives significant improvements to calculated atomization energies in numerous small molecules [53].

The geometries of $P_n$ and $As_n$ clusters change very little on using the non-local approximation [47, 48] for $E_{xc}$, and changes in the ordering of the isomer energies for a given cluster are small and restricted to clusters with more than nine atoms. The dissociation energies of $P_2$–$P_4$, however, now agree much better with the experimental values, and the wedge-shaped $P_8$ isomer is slightly less stable than two phosphorus tetrahedra, in agreement with the predictions of HF-based methods.

A further test of non-local corrections is provided by the isomers of $P_{10}$ (Fig. 5). MD/DF calculations with the LSD approximation [46] indicated that structure 5a, recognizable as a structural unit of the chains in Hittorf's phosphorus [48], was the most stable, while HF-based methods [52] favoured the $C_s$ structure 5b by a small amount (less than 0.1 eV). Incorporation of the non-local modifications to $E_{xc}$ changed the relative stability of the two isomers, with the $C_s$ form now lying $\sim 0.1$ eV lower. Some additional comments on gradient corrections to $E_{xc}$ are given in Sect. 6.

### 4.3 Group 16 Elements: S, Se

MD/DF calculations [54] performed on clusters up to $S_{13}$ showed that it is possible to determine low-lying energy minima even if the initial geometry is far from the correct structure. Starting from almost linear chains [$S_{3-6}$] or from nearly planar rings [$S_{7-13}$], the calculated structures agreed well with experiment in all cases where X-ray data were available. An example is $S_{12}$, where the symmetry ($D_{3d}$) and structural parameters (bond length $d = 3.97$ a.u., bond angle $\alpha = 106°$, dihedral angle $\gamma = 88°$) agree well with measured values

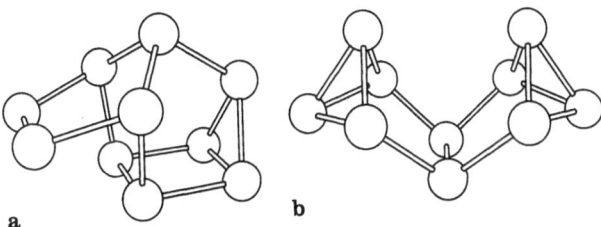

a                                                                    b

**Fig. 5a, b.** Two isomers of $P_{10}$. **a** $C_s$, **b** $C_{2v}$ [47]

(3.88 a.u., 106.2°, 87.2°, respectively). We also obtained plausible predictions in cases where single crystal specimens have not yet been prepared: In $S_5$ we found an "envelope" $(C_s)$ structure and a $C_2$ structure with almost the same energy, and the predictions of the structures of $S_4, S_5$ and $S_9$ have been confirmed by subsequent HF-based calculations [55].

The results for the $S_9$ molecule are particularly interesting. While this molecule can be prepared in microcrystalline form [11, 56], the absence of single crystals has ruled out an X-ray structure determination. However, after analyzing the Raman spectra, Steudel et al. [56] concluded that the constituent molecules have nearly identical structures, with S–S bonds in a narrow range $\sim 3.90$ a.u. From structural trends found in other sulphur ring molecules, they predicted that the dihedral angles lie in the range $70° \leq \gamma \leq 130°$, eliminating the possibility of $C_s$ symmetry and allowing only $C_1$ and $C_2$. Our calculations led to a $C_2$ ground state structure that fulfils all of the above criteria.

Perhaps the most interesting feature of the structural trends in the $S_n$ clusters is, however, the difficulty of interpolating between or extrapolating from known structures. In $S_9$, for example, the knowledge of the structures of several members of the family on either side ($n = 6$–8, 10–13), does not allow us to predict unambiguously either the structure or even the pattern of dihedral angles ["motif"] [11, 57] of the most stable isomer. Isomers with quite different structures can have very similar energies, and the assignment of the ground state requires quantitative measurements or calculations rather than qualitative arguments.

# 5 MD/DF Calculations – Photoelectron Detachment Spectroscopy

In the present section, we extend the discussion of Sect. 2.3 and show that the combination of MD/DF calculations and photoelectron detachment measurements can give useful structural information about clusters. The MD/DF calculations have a dual focus: (a) the calculation of the structures of isomers, i.e., locating the minima in the energy surfaces for each cluster anion, (b) determining, for these structures, the energy differences between the anions and states of the neutral clusters. The comparison with experiment involves both such vertical excitation energies and the vibration frequencies, quantities that are particularly useful if the values for different isomers of a cluster are distinctly different.

Energy calculations for charged systems require care in a supercell geometry with PBC, particularly where the energies of charged and neutral systems are compared. In order to calculate the VDE (ionization energies of the anions) we use the scheme of von Barth [58] to relate the multiplet energies of the neutral clusters to the energies of single determinantal states. As discussed in Ref. [59], we extrapolate the Coulomb energy of the cluster in the above unit cell to give the corresponding energy for an isolated cluster with the same charge distribution. Vibration frequencies are calculated by using an eigenmode detection scheme [60] to analyze non-thermally-equilibrated MD trajectories for the system in question. For a neutral cluster with the optimum structure of an anion, for example, we remove an electron from the anion and allow the system to evolve in MD runs (at 300–500 K) to find the closest minimum on the energy surface of the neutral cluster. The cluster atoms are then displaced by small random amounts, and the trajectories followed for 2000 to 5000 time steps at 300 K.

The negative ion source (pulsed arc cluster ion source, PACIS) has been described in detail elsewhere [61, 62], and its features are outlined here for the case of sulphur. Ions are produced in a carbon arc, the lower electrode of which is shaped like a crucible and contains a reservoir of S. A pulse of He gas is flushed through the gap between the electrodes during ignition. The He/S plasma cools in an extender and forms clusters, which grow further on cooling in a supersonic jet. Adjustments of the source that influence the structure of the clusters include the He stagnation pressure and the voltage and duration of the arc. However, the most important parameter is the time spent by the clusters in the extender.

The beam of anions is separated, according to their velocities, into a sequence of cluster bunches with a defined mass. A selected bunch is irradiated by a laser pulse of a given photon energy (2.33, 3.49 or 4.66 eV), and the electrons detached are guided by magnetic fields towards an electron detector. The kinetic energy of the electrons is related to the times-of-flight, and the binding energy (BE) is the difference between the photon energy and the kinetic energy. The energy resolution of the electron spectrometer, which depends on the kinetic energy of the electrons and the velocity and reactivity of the anions, is typically 40–70 meV.

## 5.1 Group 13 Elements

The vertical excitation energies from states of $Al_2^- \rightarrow Al_2$ can be observed in photoelectron detachment spectroscopy of negative ions, a technique that has been applied to $Al_n$ and $Ga_n$ clusters up to $n = 15$ [63]. For $Al_2^-$, there is a peak at $\sim 4.2$ eV that shows a vibrational structure with $\omega_e = 450 \pm 40$ cm$^{-1}$, in good agreement with the excitation energy from the anion (Fig. 1) and the calculated vibration frequency of the $^5\Sigma_u^-$ state of the neutral dimer (Table 1).

A detailed analysis of the structures and excitation energies in clusters of Al and Ga is in progress. We have noted above that transitions between states with the same equilibrium geometry, as is the case for the calculated ground state geometries of $Al_3^-$ and $Al_3$, should give rise to sharp peaks in the photoelectron spectrum. This is confirmed by the measurement, for which the peaks in the spectra of $Al_3^-$ and $Ga_3^-$ are the sharpest features found in these clusters [63].

## 5.2 Group 15 Elements

Structures and energies have been calculated for $P_n^-$ ions for $n = 1–9$ [64], and we focus here on aspects that are needed for the comparison with experiment. In earlier studies of the trimer anion $P_3^-$, two groups [65, 66] found three low-lying minima: an equilateral triangle $(D_{3h}, {}^3A_2')$, a linear closed-shell singlet $(D_{\infty h}, {}^1\Sigma_g^+)$, and a bent $(C_{2v})$ triplet. The first two were so close in energy that a definite prediction of the ground state was not possible. In the MD/DF calculations the linear structure has the lowest energy, but the $D_{3h}$ and $C_{2v}$ structures are only 0.06 and 0.27 eV less stable. The calculated geometries agree well with the earlier work [65, 66]. The vertical detachment energies of the three structures differ significantly, being 3.00 eV, 1.88 eV, and 1.73 eV, respectively. The inclusion of non-local modifications to the energy functional reverses the energy ordering of the two most stable non-linear structures, with energies relative to the linear structure being $-0.08$ and 0.27 eV.

The neutral phosphorus tetramer has the familiar tetrahedral $(T_d)$ structure. The additional electron in the anion, however, results in a Jahn-Teller distortion that is so large that the equilibrium structure (Fig. 6d) should not be analyzed in terms of tetrahedral symmetry. The vertical detachment energies calculated for transitions to the lowest-lying singlet and triplet structures are 1.35 and 2.91 eV, respectively. Relaxation of the neutral tetramer from the anionic structure to the tetrahedral form results in an energy lowering of 1.16 eV.

For the pentamer anion, Hartree-Fock calculations [67, 68] give a consistent picture of both geometry and vibration frequencies of the planar isomer. The MD/DF calculations lead to two low-lying isomers: The planar ring $(D_{5h}$, Fig. 6e) has bonds of length 3.96 a.u. and is 1.44 eV more stable than the structure (Fig. 6f) related to the most stable form found for the neutral pentamer. There is a striking difference between the vertical detachment energies of

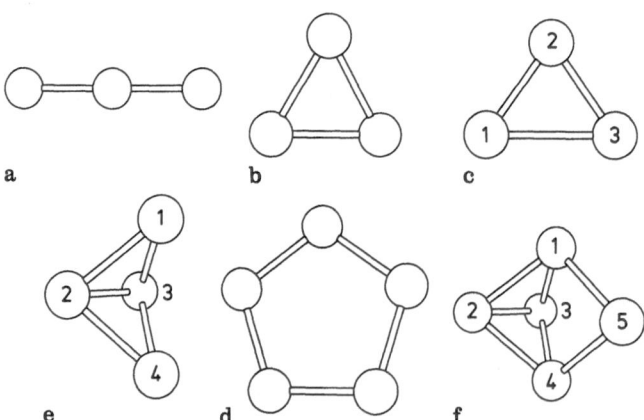

**Fig. 6a–f.** Calculated structures of **a–c** $P_3^-$, **d** $P_4^-$, and **e–f** $P_5^-$ [64]

the two isomers of the anion ($D_{5h}$: 4.04 eV, $C_{2v}$: 2.08 eV). The calculated vibration frequencies [64] are about 10% below the Hartree-Fock values, which are generally higher than experimental frequencies by approximately this amount [68].

The MD/DF calculations predict several local minima in the energy surface for $P_8^-$. The two most stable are derived from the cuneane structure, with 7a being more stable by 0.22 eV. In Fig. 7b, there is a large expansion in two of the parallel bonds. This configuration is unstable to annealing at 300 K, with one of the stretched bonds breaking and the other contracting to give structure 7a with lower symmetry ($C_s$). A similar situation occurs in the structures related to a cube, where the more symmetrical structure 7d has four expanded bonds. The more stable 7c – with energy 1.3 eV above that of 7a – has one broken bond and seven bonds of length comparable with those in the cubic form of $P_8$. Annealing from the $D_{2h}$ structure of $P_8$ [46] also results in structure 7c. The calculated VDE values are significantly higher for the most stable isomer 7a [3.0 eV] than for all others, e.g., 2.5 eV in Fig. 7c.

The trends in the structures of the phosphorus cluster anions allow those of the neutral clusters, although the anion structures are generally more open. The structures are "three-dimensional" from $n = 4$ and favour three-fold coordination, although two-fold coordinated atoms generally have shorter bonds. The shortest bonds found were the multiple bonds in the dimer and trimer. Bonds in rectangular structural units (bond angles $\sim 90°$) are generally longer than those .in triangular units, and the presence of rectangular units is energetically unfavourable.

The PACIS has been used to generate phosphorus cluster anions $P_n^-$ up to $n = 9$. Photoelectron spectra recorded at photon energies $hv = 2.33$ eV, 3.49 eV, and 4.66 eV are shown in Figs. 8, 9a, and 9b, respectively. The spectra of $P_2^-$ show four electronic transitions (Fig. 9b, A–D), and feature A is assigned to the ground state transition (VDE $0.68 \pm 0.05$ eV). Vibrational fine structure is

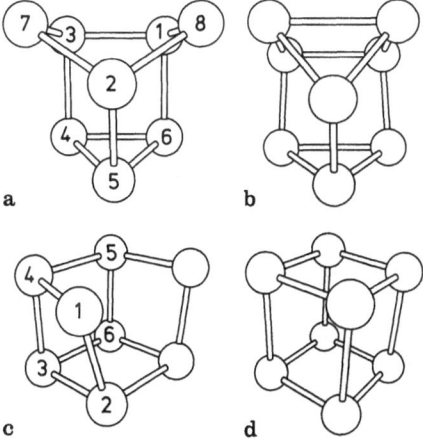

a

b

c

d

**Fig. 7a–d.** Calculated structures of $P_8^-$ [64]

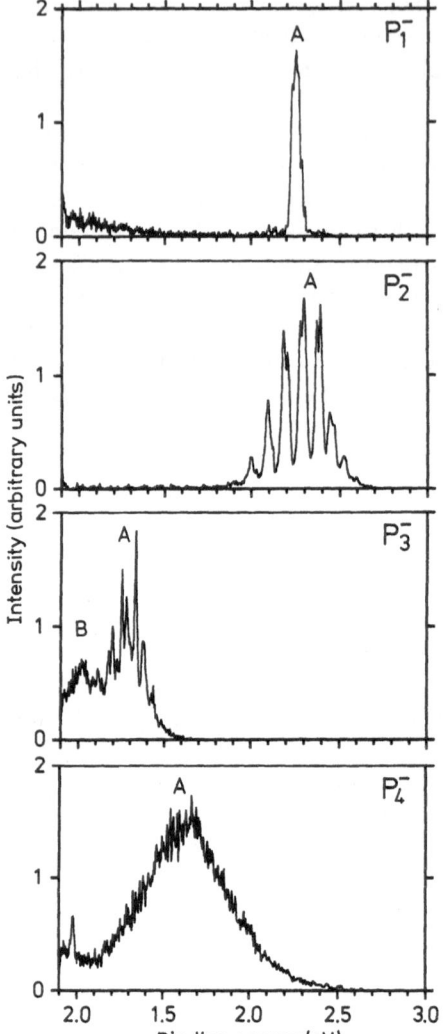

**Fig. 8.** Photoelectron spectra of $P_n^-$ clusters ($n = 1-4$) recorded at $hv = 2.33$ eV photon energy [64]

resolved at low photon energy for both dimer and trimer (Fig. 8). The calculated frequencies are in good agreement with available experimental data for $P_2$ and $P_2^-$, and the calculated values for the $a_1$ modes of the $C_{2v}$ structure of $P_3$ agree well with the measured values for this molecule.

The spectra of $P_3^-$ (Figs. 9a, b) show three peaks, and it is probable that more than one isomer can be generated, depending on the experimental conditions. The calculated vertical excitation energies to the first two states in $P_3$ are (1.73 eV, 3.70 eV for $C_{2v}$) and (1.88 eV, 4.08 eV for $D_{3h}$). Either structure could be present, but we note that *both* isomers have a large gap in the excitation spectrum, so that the strong peak at 2.89 eV must come from another structure.

Since the linear isomer gives a vertical excitation at 3.00 eV, the measured binding energy curves are consistent with the existence of at least *two* isomers, one of them linear.

There are two pronounced peaks in the measured spectra of $P_4^-$ (1.35 eV, 2.69 eV), both of which can be interpreted in terms of transitions from a "roof"-shaped isomer 6d. The calculated excitation energies to the lowest singlet and

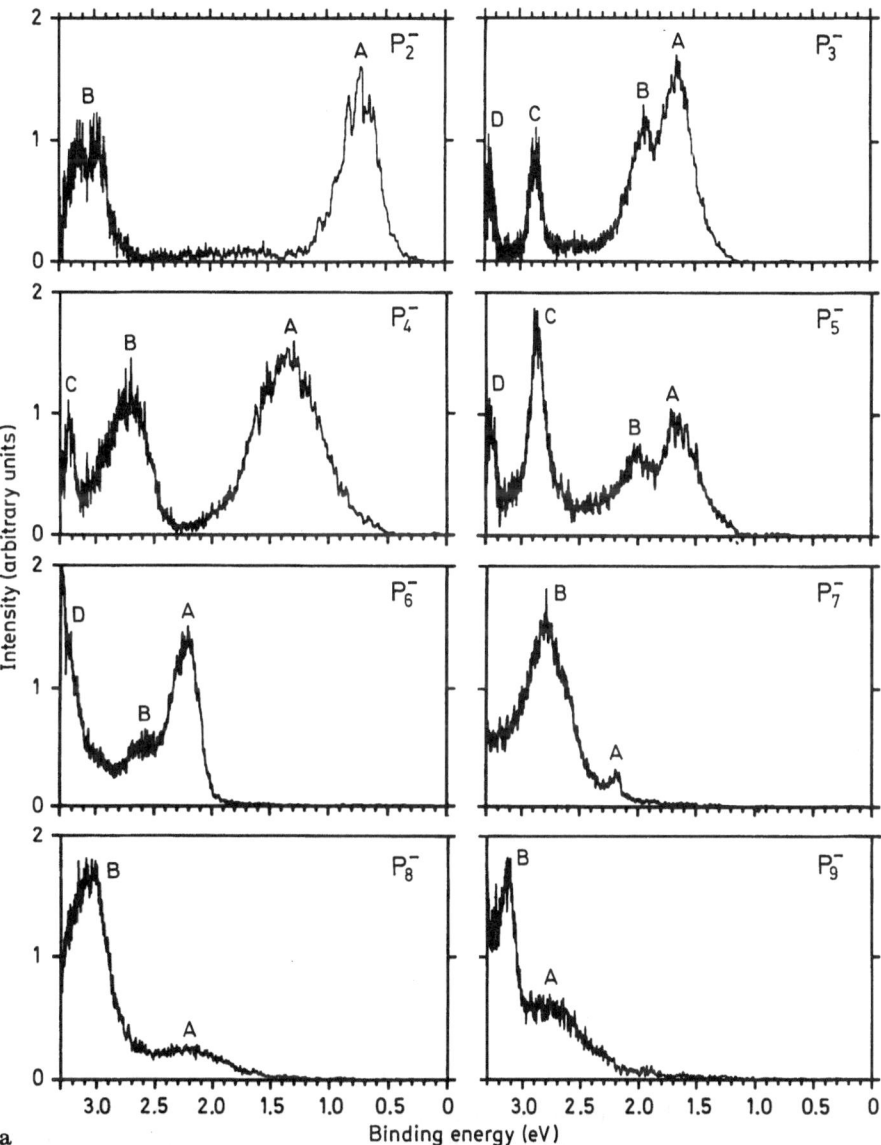

**Fig. 9a, b.** Photoelectron spectra of $P_n^-$ clusters ($n = 2$–$9$) recorded at **a** $hv = 3.49$ eV and **b** $hv = 4.66$ eV photon energies [64]

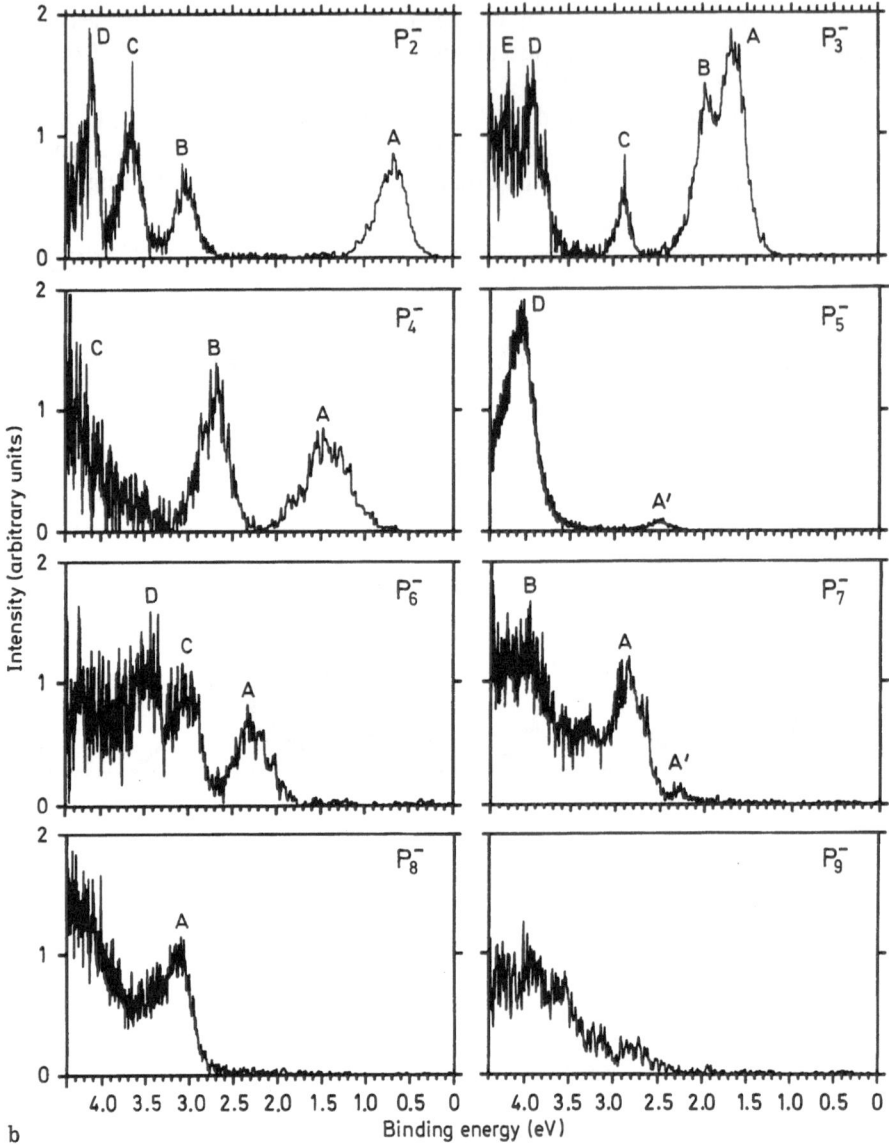

b

**Fig. 9.** *Continued*

first triplet states of $P_4$ are 1.35 eV and 2.91 eV, respectively. The broad first peak indicates a large difference between the geometries of the most stable isomers of the anion and neutral clusters. The calculated adiabatic electron affinity, the difference between the lowest energies of $P_4^-$ and $P_4$ (0.19 eV), agrees satisfactorily with the onset in the measured spectra.

The photoelectron spectra of $P_5^-$ for $hv = 3.49$ eV (9a) and 4.66 eV (9b) are very different. The features observed in the former (A, B, C) are located at almost the same BE's as the corresponding features in the spectrum of $P_3^-$ (Fig. 9a), and we assign the observed peaks to $P_3^-$ that has been generated by a photofragmentation process. The spectrum of $P_5^-$ recorded with a photon energy of 4.66 eV (9b) is consistent with a high electron affinity. A single peak at 4.04 eV dominates the spectrum of $P_5^-$ in Fig. 9b, in excellent agreement with the calculated value for the planar pentagonal form (6e). The $C_{2v}$ isomer (6f) related to the "roof plus atom" isomer predicted to be the most stable in $P_5$ lies much higher in energy and has a much lower excitation energy. There is no evidence that this isomer is generated by the PACIS.

Calculations of the vertical excitation energies for $P_8^-$ are consistent with the results of total energy calculations, which indicate that the perturbed cage structure (7a) is the most prevalent. The agreement between the measured VDE (3.05 eV) and the calculated value (3.02 eV) is very good, while the values for the other three isomers are much lower (2.34–2.55 eV).

The overall comparison between theory and experiment provides a consistent picture of the cluster isomers and their photoelectron spectroscopy. In those cases, such as $P_4^-$, $P_5^-$, and $P_8^-$, where the calculations give a definite prediction of the form of the most stable isomer, the calculated energy differences are in good agreement with experiment. The measurements indicate that the most stable form of $P_8^-$ (with high probability $P_8$ as well) has a "wedge" (cuneane) rather than a cubic structure. In $P_3^-$, where the present and earlier calculations predict the existence of three isomers with different structures but very similar energies, we show that the source generates at least two isomers, one of them linear. Spectra for $P_5^-$ taken with $hv = 3.49$ eV (below the electron affinity 4.04 eV) show fragmentation into $P_3^-$ and $P_2$.

## 5.3 Group 16 Elements

We have extended the previous MD/DF calculations on neutral sulphur clusters [54] to sulphur anions up to $S_9^-$. The results for the dimer and trimer, e.g. vibration frequencies for the anions $S_2^-$ and $S_3^-$ and for the neutral systems, agree well with previous work. Of particular interest is the prediction of stable "chain" structures in addition to the "ring" structures familiar from the neutral clusters (see above). Fig. 10 shows the most stable isomers of the former and representative isomers of the latter, of which there are many for $n > 5$. Of particular interest are chains with a planar tetramer at one or both ends, because such structures had not been found in the neutral chains. The most stable structures are generally opened or puckered rings of the neutral clusters, where at least one bond is strained or broken by the presence of the additional electron. The ground states of $S_5^-$, $S_7^-$, $S_8^-$ and $S_9^-$ belong to this family.

Of the many chainlike structures we show the all-*trans* helices in Fig. 10 (to illustrate other structures we show those of $S_7^-$ in Fig. 11). The planar motif with

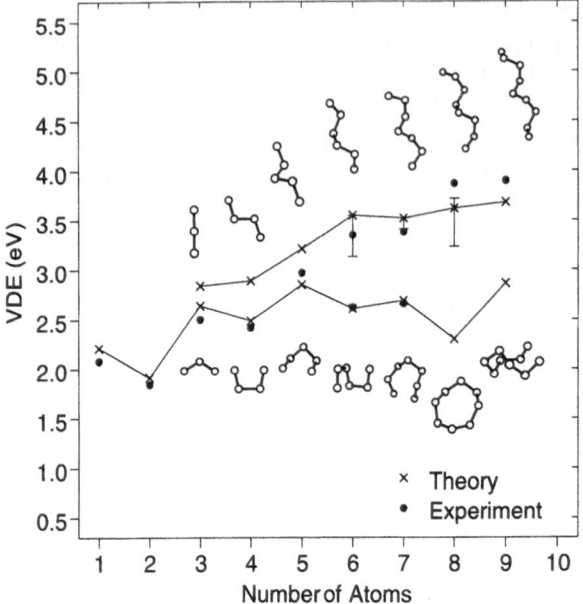

**Fig. 10.** Vertical detachment energies of sulphur anions $S_n^-$, $n = 1$–9. Circles: experiment, crosses: calculations, including values for helical chains. The bars cover the values for other chain structures [59]

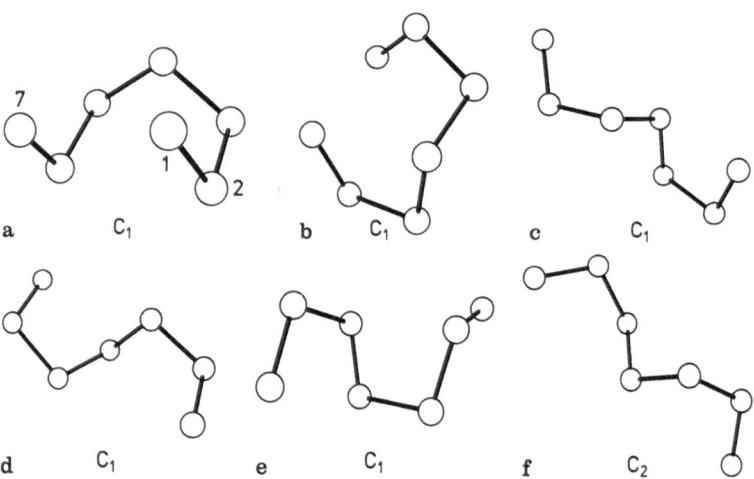

**Fig. 11a–f.** Calculated structures of $S_7^-$ [59]

approximately $C_{2v}$ symmetry occurs in $S_4^-$ and in sections of the larger anions. Structures consisting entirely of this pattern exist only for even values of $n$ ($S_4^-$, $S_6^-$, $S_8^-$), where they are among the most stable isomers. In Fig. 10 we also show the calculated VDE values for cage- and ringlike structures found for the anions. For the former, we show both the ranges of VDE values and the result for the helical chain for $n = 6$–$8$.

The VDE of the two classes of structures are strikingly different. While the broken rings show almost constant or even decreasing VDE with increasing cluster size, the values for the chains increase initially and then saturate near $S_6^-$. The outermost electron is more tightly bound in the chains, and an analysis of the densities shows that the additional electron occupies an antibonding orbital localized mainly on the terminal bonds. $S_2^-$, where the additional electron occupies an orbital with a pronounced antibonding character, has a particularly low VDE. This picture, where the potential energy is lower in chains with larger distances between the ends, is consistent with the saturation of the VDE found in longer chains.

The above predictions have been confirmed in large part by photoelectron spectroscopy of the same ions [59]. In Fig. 12 we compare mass spectra of $S_n^-$ clusters obtained at two different adjustments of the PACIS. Adjustment $a$ favours small clusters with $n = 2$–$7$, which probably result from a relatively slow "annealing" of the clusters. Apart from the trimer, clusters with $n < 6$ have a very low intensity in spectrum $b$, which is the result of more rapid cooling. The progression starts at $S_6^-$ and reaches a maximum at $S_{10}^-$. Only $S_3^-$, $S_6^-$ and $S_7^-$ have relatively high intensities in both spectra. The photoelectron spectra for

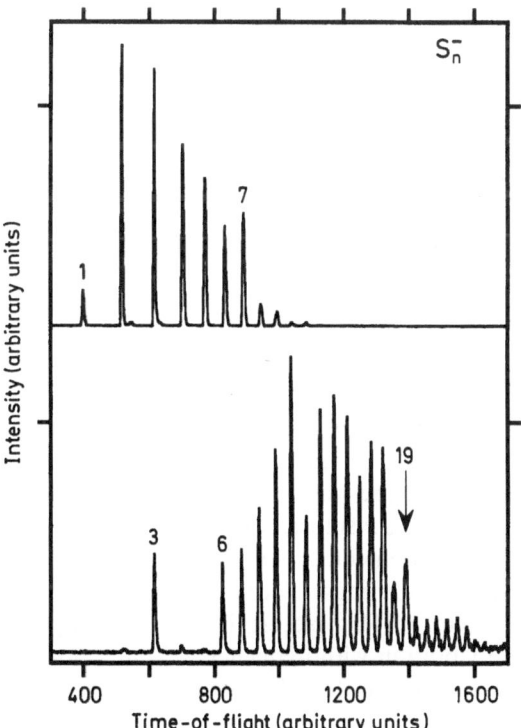

**Fig. 12.** Mass spectra of $S_n^-$ clusters. Upper frame is for adjustment $a$ (slow cooling of the sulphur plasma), lower frame for adjustment $b$ (more rapid cooling). The larger average size in the latter is due to the higher concentration of S in the carrier gas [59]

111

$S_6^-$ and $S_7^-$ (Fig. 13a, b) are quite different for the two adjustments, suggesting that different isomers have been generated.

We assign the dominant features (A, B, C) observed to photoemission from two isomers of $S_6^-$, denoted (I) and (II), respectively. The only vibrational fine structure observed for $h\nu = 4.66$ eV, feature A in the spectrum of $S_6^-$ (II), corresponds to a frequency of $570 \pm 32$ cm$^{-1}$ [64]. Two different spectra (Fig. 14c, d) have also been recorded for $S_7^-$, corresponding to two different isomers $S_7^-$ (I) and $S_7^-$ (II). We now show that the comparison between theory and experiment for the vertical detachment energies (VDE) and vibration frequencies allows us to identify both ring and chain isomers for $S_6^-$ and $S_7^-$, although no vibrational structure could be resolved in the latter. Vibrational fine structure is resolved in the $h\nu = 3.49$ eV spectra for $n = 2, 3, 4$ (Fig. 14).

Fig. 10 shows a comparison of the calculated VDE – for the most stable closed and open isomers – with values extracted from the photoelectron spectra. Values are shown for both spectra of $S_6^-$ and $S_7^-$. The overall agreement is remarkably good. For the clusters up to $S_5^-$ the experimental VDE agree with the values for the most stable closed structures to within 0.15 eV, and the two measured values for $S_6^-$ and $S_7^-$ are very close to the calculated VDE of the most stable closed and open forms. The experimental values for $S_8^-$ and $S_9^-$ are in the same range as those for the chainlike structures.

In addition to transitions to the most stable states of the neutral clusters [Fig. 10], information can be obtained by measuring transitions into excited states of $S_n$. The first three peaks in the dimer (1.84, 2.45, 2.73 eV) are in satisfactory agreement with the calculated excitation energies to the $^3\Sigma_g^-$, $^1\Delta_g$, and $^1\Sigma_g^+$ states of $S_2$ (1.91, 2.45, 2.98 eV). In $S_3^-$, the measured excitation energies (2.50, 3.7, 3.9 eV) are consistent with transitions for the open structure to the $^1A_1$, $^3A_2/^3B_1$ and $^3B_2$ states of the neutral cluster (2.64, 3.73/3.77, 3.95 eV), but not with excitations for the ring structure (1.34, 2.93, 2.94 eV for the $^1A_1$, $^3A_2$, $^3B_1$ states, respectively). The measured binding energies in $S_4^-$ are consistent with the calculated multiplet structures of the closed $C_{2v}$ and $D_{2h}$ geometries, but not with those for the open $C_{2h}$ form (2.89, 3.36, 4.07 eV for $^1A_g$, $^3B_u$, $^1B_u$, respectively).

For $S_4$, the measured frequencies are closer to the calculated values for the $C_{2v}$ and $D_{2h}$ isomers than those of the $C_{2h}$ form, which is consistent with the relative stabilities calculated. The only frequency measured for the $S_6$ structure ($570 \pm 32$ cm$^{-1}$) is significantly higher than both the calculated and Raman frequencies of the $D_{3d}$ isomer and falls in a pronounced gap of the spectrum for the $D_{3h}$ isomer. Since the vibrational structure was only observed in spectrum $S_6^-$ (II), it appears that the neutral cluster has insufficient time to relax to one of the more stable isomers during the measurement. It is also consistent with the observation [69] that *all* unbranched sulphur rings regardless of the size have no fundamental frequencies above 530 cm$^{-1}$. This picture is supported by the existence of a totally symmetric ($a$) vibration with frequency 619 cm$^{-1}$ for the neutral hexamer with the helical ($C_2$) geometry found for the $S_6^-$ anion.

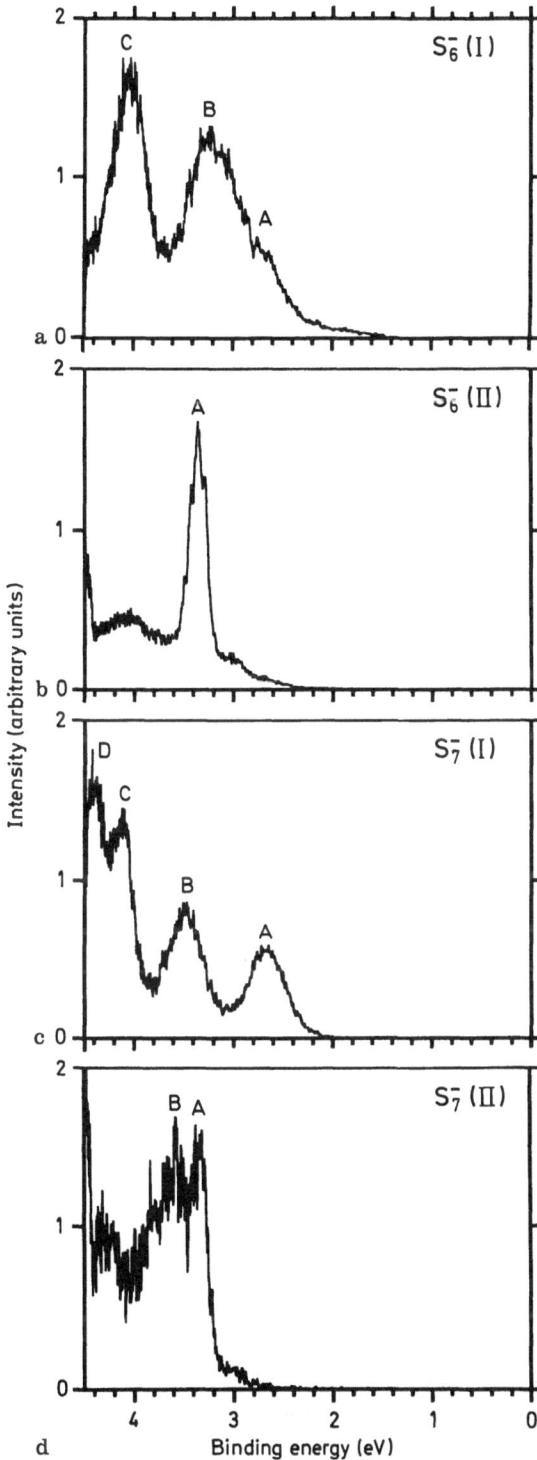

**Fig. 13a–d.** Photoelectron spectra of $S_6^-$ and $S_7^-$ recorded at $h\nu = 4.66$ eV. Spectra are shown for the two adjustments corresponding to the mass spectra in Fig. 12 [59]

The comparison between theory and experiment (Fig. 10) indicates that clusters generated by the source are ringlike up to $S_5^-$ and chainlike for $S_8^-$ and $S_9^-$. $S_6^-$ and $S_7^-$ can occur in both forms, with source adjustments $a$ and

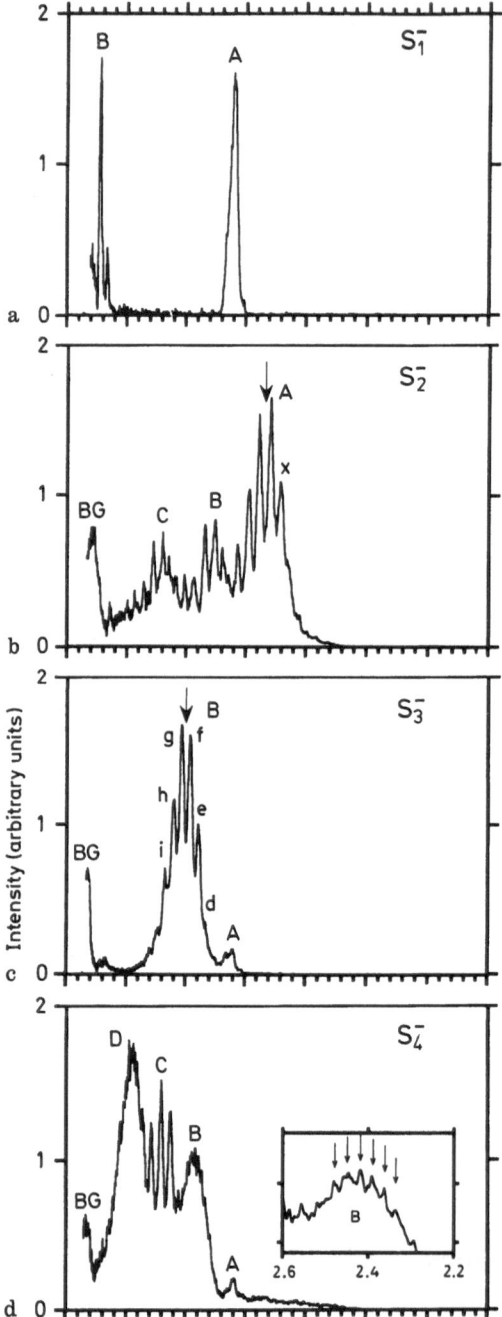

**Fig. 14a–e.** Photoelectron spectra of $S_n^-$ clusters $(n = 1–5)$ recorded for $h\nu = 3.39$ eV photon energy [59]

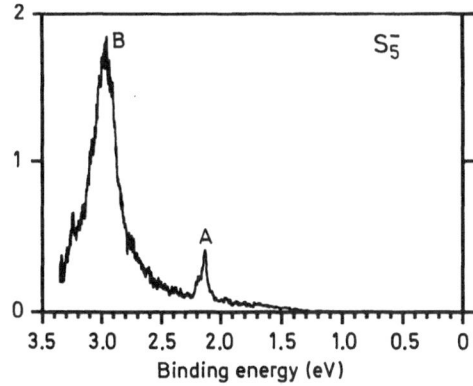

**Fig. 14.** *Continued*

*b* favouring rings (low VDE) and chains (higher VDE), respectively. The photo-electron spectra [Fig. 14a–d] then provide spectroscopic evidence for the exist-ence of different cluster isomers in the gas phase. The calculated and measured vibration frequencies are consistent with this picture.

The transition from closed to open structures as *n* increases may be surpris-ing at first sight, since our calculations predict that closed isomers are the most stable for *all* cluster sizes. In fact, the smallest clusters have few isomers, the relaxation of the structure is more rapid, and only the most stable ringlike structures are observed. A relatively low energy is, however, not the sole criterion for the occurrence of particular structures in a cluster beam. Rings are observed for $S_6^-$ and $S_7^-$ when conditions allow a slower cooling of the plasma and more time for structural rearrangement. Structure formation is necessarily more complex in closed structures than in chains, where growth can occur by the addition of terminal atoms. Ring formation from $S_n^-$ chains is hampered by the negative charge localized on the terminal atoms, in contrast to neutral clusters, where simulated annealing indicates that all chain structures relax to a closed form if the additional electron is removed.

The preference for chain isomers in the larger clusters can be appreciated by examining the pattern of the signs of the dihedral angles ("motif") in different structures. Closed structures require distinct patterns (such as $+ - + - + - + -$ in the most stable isomer of $S_8^-$), while open structures can have many combinations of $+$, $-$, and 0, since the beginning and the end of the chain are not constrained to coincide. The configurational freedom of chains means that they are favoured if the time available for cluster formation is too short to allow annealing. The observation of chainlike isomers is consistent with the higher electron affinities found in these structures [59], since such clusters are more likely to survive growth, fragmentation and charge transfer processes. Clusters with lower EA are not excluded, as we have seen in $S_6^-$ and $S_7^-$, and suitable annealing of the cluster beam could lead to ringlike structures.

# 6 Discussion and Concluding Remarks

The geometrical structure of molecules and clusters is a topic of widespread importance. For the theorist this involves two distinct aspects: the calculation of the total energy of the system of electrons and ions for a given geometry, and the determination of the most stable of the structures associated with the many minima in the energy surface.

The approach we have discussed here addresses both problems with comparable emphasis. The density functional formalism, with the LSD approximation for the exchange-correlation energy, provides us with an approximate method of calculating energy surfaces, and the results have predictive value in many contexts. DF can also be carried out with comparable ease for all elements. When coupled with MD at elevated temperatures (simulated annealing), it is possible to study cases where the most stable isomers are unknown, or where the energy surfaces have many local minima.

The acceptance of DF methods as a method for calculations on clusters and small molecules has been enhanced by the improved results obtained for atomization energies when non-local approximations for $E_{xc}$ are used. Gradient corrections give, for example, the same ordering of levels in $P_{10}$ and the same relative stability of $P_8$ and $2P_4$ that were found in HF-based calculations. It would be incorrect, however, to assume that non-local corrections provide a panacea for problems that may arise in using the LSD approximation. Recent work in isomers of $C_{20}$ [70], for example, has shown that the inclusion of gradient corrections changes the relative stability of cage and ring isomers by more than 7 eV. There are also substantial deviations from the experimental multiplet structure of $C_2$. Although gradient corrections yield very good agreement with experiment for the formation energy of the measured ground state ($^1\Sigma_g^+$), the next highest state ($^3\Pi_u$) is predicted to be $\sim 0.7$ eV *more* stable, a relative error that is larger than in LSD calculations. There remains considerable scope for the improvement of more reliable energy functionals in the DF framework. In any event, the geometries found in LSD calculations generally provide very reliable estimates that can be used as input for ab initio methods of energy calculation, as done in the QMC calculations of Ref. [31].

The combination of MD/DF calculations and photodetachment spectroscopy provides a useful approach to the problem of determining the structure of atomic clusters. The PACIS provides a flexible method for generating clusters of many materials and for performing photoelectron spectroscopy on them. The elements include metals (solid and liquid), semiconductors, and highly reactive materials such as phosphorus. There is little doubt that the source will provide interesting data for many other systems. For the analysis of these data, or for predictions of interesting systems for study, the MD/DF approach should prove to be an ideal tool.

*Acknowledgements.* I thank numerous colleagues in Jülich, particularly P. Ballone, G. Ganteför, and S. Hunsicker, for discussions and collaboration on work described here. The calculations were performed on Cray computers of the Forschungszentrum Jülich and the German Supercomputer Centre (HLRZ).

# 7 References

1. The "International Symposium on Small Particles and Inorganic Clusters" takes place biennially. The Proceedings of the Sixth Symposium (Chicago, 1992) are published in Z. Phys. D – Atoms, Molecules and Clusters, Vol. 26. The Proceedings of the Seventh Symposium (Kobe, 1994) are in press, and the Eighth Symposium will be held in Copenhagen in 1996
2. See, for example, Scoles G (ed) (1990) The Chemical Physics of Atomic and Molecular Clusters, Proc. Scuola Internazionale di Fisica, Course CVII. North-Holland, Amsterdam
3. See, for example, Haberland H (ed) (1994) Clusters of Atoms and Molecules. Springer, Berlin Heidelberg New York
4. Jones RO (1990) in: Campagna M, Rosei R (eds) Photoemission and absorption Spectroscopy of Solids and Interfaces with Synchrotron Radiation. Proc. Scuola Internazionale di Fisica, Course CVIII. North-Holland, Amsterdam
5. Jones RO (1994) J. Molecular Struc. (Theochem) 308: 219
6. Crick F (1988) in: What mad pursuit. Penguin, London, p 150
7. Kirkpatrick S, Gelatt CD, Vecchi MP (1983) Science 220: 671
8. Jones RO, Gunnarsson O (1989) Rev Mod Phys 61: 689
9. Car R, Parrinello M (1985) Phys Rev Lett 55: 2471
10. Donohue J (1974) The Structures of the Elements, Wiley, New York, Chapters 5 [group 13], 8 [group 15] and 9 [group 16]
11. Steudel R (1984) in: Studies in Inorganic Chemistry, Vol. 5, Müller A, Krebs B (eds) Elsevier, Amsterdam
12. Steudel R, Strauss EM (1987) in: The Chemistry of Inorganic Homo- and Heterocycles, Vol. 2, Academic, London, p 769
13. Bitterer H (ed) (1980) Schwefel: Gmelin Handbuch der Anorganischen Chemie, 8. Aufl., Ergänzungsband 3. Springer, Berlin, p 8
14. Huber KP, Herzberg G (1979) Molecular Spectra and Molecular Structure. IV. Constants of Diatomic Molecules. Van Nostrand Reinhold, New York
15. Brassington NJ, Edwards HGM, Long DA (1981) J Raman Spectrosc 11: 346
16. Honea EC, Ogura A, Murray CA, Raghavachari K, Sprenger WO, Jarrold MF, Brown WL (1993) Nature (London) 366: 42
17. Raghavachari K, Rohlfing CM (1988) J Chem Phys 89: 2219
18. Martin TP (1984) J Chem Phys 81: 4427
19. Martin TP (1986) Z Phys D 3: 221. These studies have recently been extended to $P_n^+$ clusters with $n > 6000$ (Martin TP, Näher U, private communication)
20. Zimmerman JA, Bach SBH, Watson CH, Eyler JR (1991) J Phys Chem 95: 98
21. Wolf S, Sommerer G, Rutz S, Schreiber E, Leisner T, Wöste L, Berry RS (1995) Phys. Rev. Lett. 74: 4177
22. See, for example, Mead RD, Stephens AE, Lineberger WC (1984) in: Bowers MT (ed) Gas Phase Ion Chemistry. Academic, Orlando, p 213 and references therein
23. Ho J, Ervin KM, Lineberger WC (1990) J Chem Phys 93: 6987
24. McHugh KM, Eaton JG, Lee GH, Sarkas HW, Kidder LH, Snodgrass JT, Manaa MR, Bowen KH (1989) J Chem Phys 91: 3792
25. Eaton JG, Kidder LH, Sarkas HW, McHugh KM, Bowen KH (1991) in: Schmidt R, Lutz HO, Dreizler R (eds), Nuclear Physics Concepts in the Study of Atomic Cluster Physics. Springer, Berlin Heidelberg, p 291
26. Yang S, Taylor KJ, Craycraft MJ, Conceicao J, Pettiette CL, Cheshnovsky O, Smalley RE (1988) Chem Phys Lett 144: 431
27. Polak ML, Ho J, Gerber G, Lineberger WC (1992) J Chem Phys 95: 3053 [$Bi_n$]; Polak ML, Gerber G, Ho J, Lineberger WC (1992) J Chem Phys 97: 8990 [$Sb_n$]

28. Arnold CC, Neumark DM (1993) J Chem Phys 99: 3353
29. Rohlfing CM, Raghavachari K (1992) J Chem Phys 96: 2114
30. Bonačić-Koutecký V, Fantucci P, Koutecký J (1991) Chem Rev 91: 1035
31. Grossman JC, Mitáš L (1995) Phys Rev Lett 74: 1323
32. Hohenberg P, Kohn W (1964) Phys Rev 136: B864
33. Kohn W, Sham LJ (1965) Phys Rev 140: A1133
34. Bachelet GB, Hamann DR, Schlüter M (1982) Phys Rev B 26: 4199
35. Stumpf R, Gonze X, Scheffler M (1990) Research Report, Fritz-Haber-Institut, Berlin, unpublished
36. Hohl D, Jones RO, Car R, Parrinello M (1989) J Am Chem Soc 111: 825
37. Cox DM, Trevor DJ, Whetten RL, Rohlfing EA, Kaldor A (1986) J Chem Phys 84: 4651 [$n = 2$–25]
38. Cox DM, Trevor DJ, Whetten RL, Kaldor A (1988) J Phys Chem 92: 421 [$n = 2$–13]
39. de Heer WA, Milani P, Châtelain A (1989) Phys Rev Lett 63: 2834 [up to $n = 61$]
40. (a) Jarrold MF, Bower JE, Kraus JS (1987) J Chem Phys 86: 3876 [$n = 3$–26];
    (b) Hanley L, Ruatta SA, Anderson SL (1987) J Chem Phys 87: 260 [$n = 2$–7]
41. Cai MF, Djugan TP, Bondybey VE (1989) Chem Phys Lett 155: 430
42. Sunil KK, Jordan KD (1988) J Phys Chem 92: 2774
43. Bauschlicher CW Jr, Partridge H, Langhoff, Taylor PR, Walch SP (1987) J Chem Phys 86: 7007
44. Jones RO (1991) Phys Rev Lett 67: 224; Jones RO (1993) J Chem Phys 99: 1194
45. Corbridge DEC (1985) Phosphorus. An Outline of its Chemistry, Biochemistry and Technology. Elsevier, Amsterdam
46. Jones RO, Hohl D (1990) J Chem Phys 92: 6710; Jones RO, Seifert G (1992) J Chem Phys 96: 7564
47. Ballone P, Jones RO (1994) J Chem Phys 100: 4941
48. Thurn H, Krebs H (1969) Acta Cryst B 25: 125
49. Eaton PE, Cole TW Jr (1964) J Am Chem Soc 86: 962, 3157
50. Cassar L, Eaton PE, Halpern J (1970) J Am Chem Soc 92: 6366
51. Janoschek R (1992) Chem Ber 125: 2687
52. Häser M, Schneider U, Ahlrichs R (1992) J Am Chem Soc 114: 9551
53. See, for example, Becke AD (1992) J Chem Phys 96: 2155; Johnson BG, Gill PMW, Pople JA (1992) J Chem Phys 97: 7846
54. Hohl D, Jones RO, Car R, Parrinello M (1988) J Chem Phys 89: 6823
55. Raghavachari K, Rohlfing CM, Binkley JS (1990) J Chem Phys 93: 5862
56. Steudel R, Sandow T, Steidel J (1985) Z. Naturforsch Teil B 40: 594
57. Tuinstra F (1967) Structural Aspects of the Allotropy of Sulphur and Other Divalent Elements. Waltman, Delft
58. von Barth U (1979) Phys Rev A 20: 1693
59. Hunsicker S, Jones RO, Ganteför G (1995) J Chem Phys 102: 5917
60. Margl P, Schwarz K, Blöchl P (1994) J Chem Phys 100: 8194
61. Cha CY, Ganteför G, Eberhardt W (1992) Rev Sci Instrum 63: 5661
62. (a) Ganteför G, Siekmann HR, Lutz HO, Meiwes-Broer KH (1990) Chem Phys Lett 165: 293;
    (b) Siekmann HR, Lüder C, Faehrmann J, Lutz HO, Meiwes-Broer KH (1991) Z Phys D 20: 417
63. Cha CY, Ganteför G, Eberhardt W (1994) J Chem Phys 100: 995
64. Jones RO, Ganteför G, Hunsicker S, Pieperhoff P (1995) J Chem Phys 103: 9549
65. Burdett JK, Marsden CJ (1988) New J Chem 12: 797
66. Hamilton TP, Schaefer HF III (1990) Chem Phys Lett 166: 303
67. Hamilton TP, Schaefer HF III (1989) Angew Chem 101: 500
68. Janoschek R (1989) Chem Ber 122: 2121
69. Lenain P, Picquenard E, Corset J, Jensen D, Steudel R (1988) Ber Bunsenges. Phys Chem 92: 859
70. Raghavachari K, Strout DL, Odom GK, Scuseria GE, Pople JA, Johnson BG, Gill PMW (1993) Chem Phys Lett 214: 357

# Density Functional Theory of Clusters of Nontransition Metals Using Simple Models

J.A. Alonso and L.C. Balbás

Departamento de Física Teórica, Universidad de Valladolid, 47011 Valladolid, Spain

## Table of Contents

Topics in Current Chemistry, Vol. 182
© Springer-Verlag Berlin Heidelberg 1996

The application of the density functional formalism to the analysis of the electronic structure of clusters of nontransiton metals, in particular alkali metals, is reviewed. The emphasis is on simple models that can be applied to medium-size and large clusters: spherical jellium model (SJM), deformed jellium model (DJM) spherically averaged pseudopotential model (SAPS) and cylindrically averaged pseudopotential model (CAPS). The main characteristic of this class of clusters is the formation of electronic shells, whose effects are manifested in the peculiar variation of cluster stability as a function of size. These shell effects persist up to very large cluster sizes. They also seem to control the dissociation behavior of multiply charged species. The second main topic is the response of small clusters to an external (dipole) electric field. At the appropriate frequency the whole valence-electron cloud responds, executing collective oscillations against the ionic background. We also review the effect of impurities, as well as the effects of mixing and segregation in clusters formed by two or three different elements. Finally, clusters of noble metals are briefly discussed.

# 1 Introduction

The terms *cluster* or *small particle* are used to denote an aggregate containing from a few atoms up to a few thousand atoms. Due to their small size, the properties of the clusters are often different from those of the corresponding bulk material. By studying the behavior of clusters, one expects to obtain information on the early stages of the growth of matter. Typical questions which arise are: How many atoms a cluster of a metallic element must contain in order to develop metallic properties?, or how properties like geometric structure and melting temperature change with cluster size? These and similar questions have motivated the development of experimental techniques for producing small clusters, as well as a series of experimental and theoretical studies of their properties. The key to the explanation of most properties of clusters is the large surface to volume ratio.

Supported bimetallic clusters are used as catalysts for the conversion of automobile exhausts to nontoxic gases and the refinement of crude oil in the petroleum industry [1]. Great expectations exist for the synthesis of exotic materials, an example being the fullerite crystal, in which the units are $C_{60}$ clusters [2]. The miniaturization of electronic components may soon reach sizes at which cluster physics becomes relevant. The existence of particularly stable clusters is often advocated in the construction of models of amorphous systems. These examples provide evidence for the technological importance of small or medium-size atomic clusters.

One of the theoretical techniques that has been used most fruitfully to explain the experimental observations about cluster properties is *density functional theory* (DFT) [3,4]. In this paper we review that work, with emphasis on the simple models that have led to a unified view of many cluster properties. For this reason, we restrict the discussion to simple metallic elements, in particular (although not exclusively) to alkali metals, for which such simple models best apply.

# 2 Types of Clusters

Clusters can be classified according to the type of chemical bonding between the atoms forming the aggregate. Interactions between inert-gas atoms involve a weak central pair force. This means that *Van der Waals clusters* will be characterised by close-packing of atoms. In *Metallic clusters* the interatomic forces are more complex and, in some cases, partly directional. The simple metallic elements (Na, Al, etc) reveal a non-smooth variation of their properties as a function of cluster size. On the other hand, the variation of the properties of transition metal clusters, although less spectacular, is interesting because of their catalytic applications. Covalent bonding is the dominating factor leading to *network clusters* of materials like Si, Ge and C; a well known example is $C_{60}$. The *clusters of Ionic Materials* like NaCl or CuBr are composed of closed-shell positive and negative ions. *Molecular clusters* are typical of organic molecules and of some closed-shell molecules like $I_2$. Finally, *hydrogen-bonded clusters* are formed by closed-shell molecules containing $H$ and electronegative elements.

Since our interest is in the applications of DFT, we will concentrate on metallic clusters, which provide ideal systems on which to apply this formalism.

121

# 3 Magic Numbers of Alkali Metal Clusters

Sodium vapour, or other alkaline vapours, can be expanded supersonically from a hot stainless steel oven with a fine exit nozzle, resulting in well focussed cluster beams. Clusters form as a result of collisions between Na atoms in the tiny expansion zone, terminating some tenths of a millimeter beyond the nozzle. The clusters warm up because the condensation is an exothermic reaction, so there also is a tendency for evaporation from the clusters. As the expansion proceeds, collisions between Na atoms end and the tendency of atoms to evaporate from the hot clusters dominates. Each cluster loses mass and cools down. In the evaporation chains, clusters with low evaporation rates, i.e., with strong binding energies, tend to become abundant.

In 1984, Knight and his coworkers performed an experiment of this kind [5,6]. They found an abundance distribution showing prominent maxima and/or steps at cluster sizes $N = 8, 20, 40, 58$ and 92. The arguments given above indicate that clusters composed of $8, 20, 40, 58$ and 92 atoms are especially stable. Since Na is a monovalent atom, the total number of valence electrons in these clusters is also 8, 20, 40, 58 and 92, respectively. Similar experiments have confirmed the same magic numbers in the mass spectra of other alkaline elements (Li, K, Rb, and Cs). Furthermore, measurements of the ionization potential, IP, as a function of cluster size show that the value of IP drops abruptly between $N$ and $N + 1$ at precisely the values $N = 8, 20, 40, 58$ and 92, that is, at the magic numbers, as well as at $N = 2$ and $N = 18$ [6]. This result shows that the electrons are bound more tightly in the magic clusters.

Cluster stabilities can also be deduced from dissociation energies in fragmentation experiments [7]. In a typical photodissociation experiment, cluster ions like $Na_N^+$ are excited by laser light to a highly excited state $(Na_N^+)^*$. The excited cluster can evaporate a neutral atom,

$$(Na_N^+)^* \rightarrow Na_{N-1}^+ + Na \tag{1}$$

if enough excitation energy to overcome the binding energy D of the atom is localized into a single vibrational mode:

$$D = E(Na_{N-1}^+) + E(Na) - E(Na_N^+) > 0 . \tag{2}$$

Statistical methods, together with experimental information on the fraction of dissociated clusters, have been used by Bréchignac et al. [7] to obtain the binding energy, D, in the case of the photodissociation of $Na_N^+$ and $K_N^+$. The most relevant conclusion is the occurrence of abrupt drops of the evaporation energy between $Na_3^+$ and $Na_4^+$, between $Na_9^+$ and $Na_{10}^+$ and between $Na_{21}^+$ and $Na_{22}^+$; similar behavior is observed in the case of K. The photodissociation experiments are performed for ionized clusters, in which the number of valence electrons is $N_e = N - 1$. Thus, high binding energies occur for clusters with 2, 8 and 20 electrons. The dissociation experiments indicate unambiguously that the magic character is associated with the number of valence electrons in the cluster, and corroborate the magic numbers obtained in the abundance spectra. The same conclusion can evidently be deduced from the ionization potentials or from the analysis of the mass spectra of clusters generated by the liquid metal ion source (LMIS) technique [8], in which nascent cluster ions are produced directly.

# 4 Spherical Jellium Model

Solid state physicists are familiar with the free- and nearly free-electron models of simple metals [9]. The essence of those models is the fact that the effective potential seen by the conduction electrons in metals like Na, K, etc., is nearly constant through the volume of the metal. This is so because: (a) the ion cores occupy only a small fraction of the atomic volume, and (b) the effective ionic potential is weak. Under these circumstances, a constant potential in the interior of the metal is a good approximation—even better if the metal is liquid. However, electrons cannot escape from the metal spontaneously: in fact, the energy needed to extract one electron through the surface is called the work function. This means that the potential rises abruptly at the surface of the metal. If the piece of metal has microscopic dimensions and we assume for simplicity its form to be spherical – like a classical liquid drop, then the effective potential confining the valence electrons will be spherically symmetric, with a form intermediate between an isotropic harmonic oscillator and a square well [10]. These simple model potentials can already give an idea of the reason for the magic numbers: the formation of electronic shells.

In general, energy levels for electrons bound in a spherically symmetric potential are characterised by radial and angular-momentum quantum numbers, $k$ and $l$ respectively ($k$-1 is equal to the number of nodes in the radial wave function.) For fixed $k$ and $l$, the magnetic quantum number $m$ can take the values $m = l, l$-1, ... -$l$, and the spin quantum number takes the two values $s = +1/2$ and $-1/2$. This gives the total degeneracy $2(2l + 1)$ for a $(k, l)$ subshell. We know from atomic and molecular physics that closed-shell configurations are very stable, because of the large energy gaps between electronic shells. However, the detailed form of the confining potential controls the precise relative ordering of the $(k, l)$ subshells, and also dictates which gaps are large and which are small where only large gaps lead to enhanced stability. So, for a precise explanation of the magic numbers of the alkali-metal clusters, a realistic representation of the effective potential is required.

An accurate selfconsistent potential can be constructed by applying DFT [3, 4, 11] within the context of the *spherical jellium model* (SJM) [12]. In this model the background of positive ions is smeared out over the volume of the cluster, to form a positive charge distribution with density:

$$n_+(r) = \begin{cases} n^o_+, & r < R \\ 0, & r < R \end{cases}.$$  (3)

The radius $R$ of the background is related to the number of atoms $N$ in the cluster by the equation,

$$\frac{4}{3}\pi R^3 = N\Omega$$  (4)

where $\Omega$ is the experimental volume per atom in the bulk metal. The constant $n^o_+$ is related to $\Omega$ and to the valence $Z(Z = 1$ for alkali elements) by

$$Z = n^o_+ \Omega.$$  (5)

This positive background provides the external attractive potential. (Hartree atomic units will be used through the paper unless explicitly stated):

$$V_{ext}(r) = -\int \frac{n_+(r')}{|r - r'|} d^3r'$$  (6)

which is parabolic inside the sphere of radius $R$, and purely coulombic outside.

DFT is then used to calculate the ground state electronic distribution for interacting electrons in this external potential. This is achieved by solving the Kohn–Sham single-particle equations,

$$\left(-\frac{1}{2}\nabla^2 + V_{eff}(r)\right)\phi_i(r) = \varepsilon_i\phi_i(r)$$  (7)

and constructing the electron density from the occupied single particle orbitals:

$$n(r) = \sum_{i=1}^{occ} |\phi_i(r)|^2.$$  (8)

The effective potential in Eq. (7) represents the average effect of the attraction from the ions and the repulsion from the other electrons. It is given by

$$V_{eff} = V_{ext} + V_H + V_{xc}.$$  (9)

$V_H$ is the classical repulsive electrostatic potential of the electronic cloud,

$$V_H(r) = \int \frac{n(r')}{|r - r'|} d^3r'$$  (10)

and $V_{xc}$ is the exchange and correlation part. Normally $V_{xc}(r)$ is calculated using the local density approximation (LDA) [4].

The spherical jellium model has been applied to alkali metal clusters by many authors (see Ref. [6]). Fig. (1) shows the self-consistent effective potential for a sodium cluster with twenty atoms. The degenerate levels are filled up to electron number $N_e = 20$. In a spherical cluster with 21 electrons, the last electron will have to occupy the $1f$ level above (dashed line). This electron is less

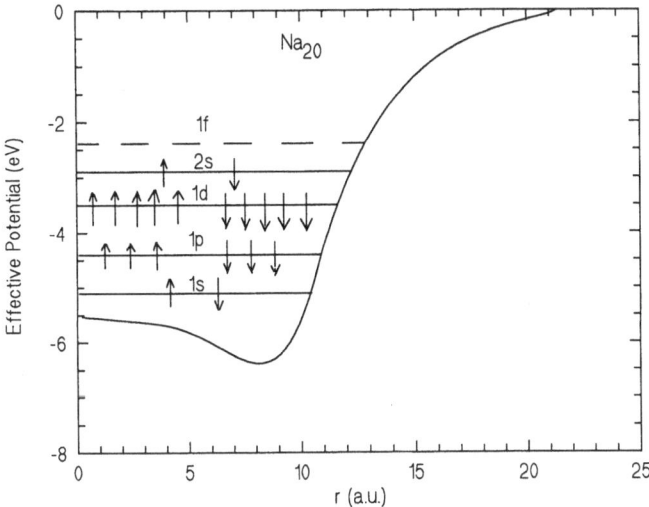

**Fig. 1.** Self-consistent effective potential for $Na_{20}$ in the spherical jellium model. The occupied electronic shells are indicated, as well as the lowest unoccupied shell.

firmly bound than the 20th electron by at least 0.5 eV, and should be easier to remove by photo-ionization. This explains why the ionization potential drops with the opening of a new shell. The total energy of the cluster, $E(N)$, can also be calculated as usual in DFT:

$$E(N) = \sum_{i=1}^{occ} -\frac{1}{2} \langle \phi_i | \nabla^2 | \phi_i \rangle$$

$$+ \frac{1}{2} \iint \frac{n(r)n(r')}{|r-r'|} \, d^3r \, d^3r' + \int V_{ext}(r)n(r) \, d^3r + E_{xc} + E_{self} . \quad (11)$$

The first term in this expression is the kinetic energy of the electrons, the second is their classical electrostatic interaction, and the third gives the interaction between the electronic cloud and the positive background. $E_{xc}$ is the exchange-correlation energy of the electrons and, finally, $E_{self}$ is the electrostatic self-interaction of the positive background.

The total energy per atom, $E(N)/N$, of alkali metal clusters in this model is a smooth function of $N$ except for kinks at $N = 8, 18, 20, 34, 40, 58, 92, ...$ [6]. To better display the abrupt changes in the total energy, we can define a quantity

$$\Delta_2(N) = E(N+1) + E(N-1) - 2E(N) \quad (12)$$

which represents the relative stability of a cluster with $N$ atoms in comparison to clusters with $(N+1)$ and $(N-1)$ atoms. If the highest occupied level is just filled by the electrons in a cluster of $N$ atoms, and the next available level is separated from this filled level by a sizable energy gap, the total cluster energy will jump

125

from $E(N)$ to $E(N + 1)$. This gives rise to a peak in $\Delta_2(N)$. A peak in $\Delta_2(N)$ then indicates that a cluster of size $N$ is very stable. The higher stability suggests that this cluster should be more abundant in the mass spectrum than clusters with $N + 1$ or $N - 1$ atoms. $\Delta_2(N)$ is shown in Fig. 2 for lithium, sodium, and potassium clusters with $N$ up to 95. Peaks in $\Delta_2(N)$ appear at $N = 2$ (not shown), 8, 18, 20, 34, 40, 58 and 92. This is consistent with the experimental mass spectra discussed above.

The calculation confirms that the magic numbers are due to the closing of electronic shells: The levels are filled in the order $1s$, $1p$, $1d$, $2s$, $1f$, $2p$, $1g$, $2d$, $1h$, $3s$... Filling these levels with the maximum number of electrons allowed leads to the subshell closing numbers 2, 8, 18, 20, 34, 40, 58, 68, 90, 92,... The number $N = 34$, which appears after filling the $1f$ level is a magic number of secondary importance, that is also observed in the experiments. On the other hand the

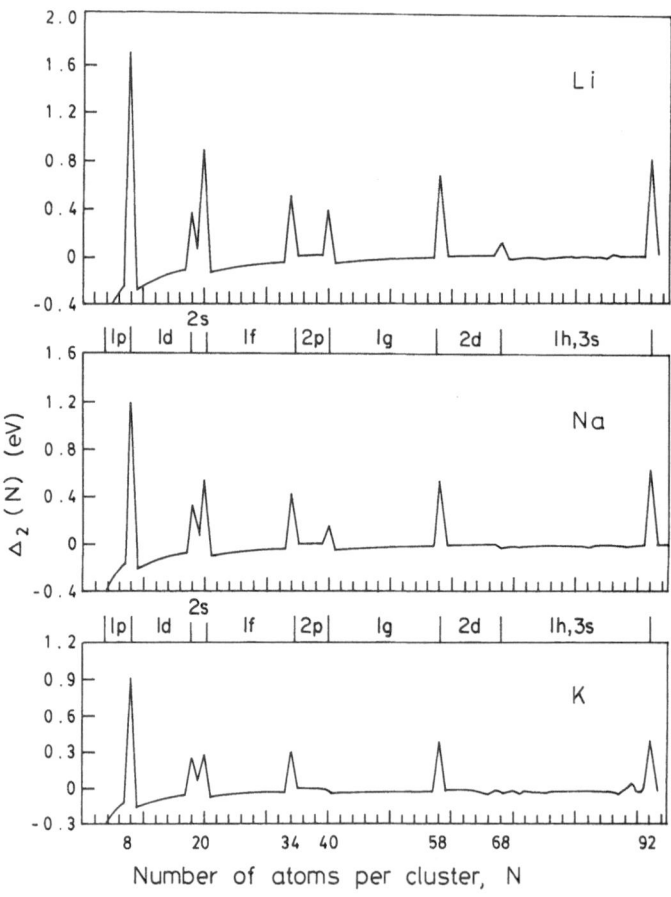

**Fig. 2.** Relative stability $\Delta_2(N) = E(N + 1) + E(N - 1) - 2E(N)$ as a function of cluster of size in the spherical jellium model.

numbers 68 and 90, which correspond to the filling of the *2d* and *1h* shells, are more difficult to observe, because the gaps between the *2d* and *1h* levels, and between the *1h* and *3s* levels, are small. Groups of levels that are close in energy will be called shells, so only the gaps between shells lead to observable consequences.

The calculated ionization potentials of Li, Na and K reproduce the drops associated with the closing of electronic shells [6]. However, the spherical jellium yields sawtoothed curves which lack fine structure between shell closings. In addition, the sawtooth rises above the experimental data before falling sharply at the next shell-closure. This behavior contrasts with the observed ionization potential curves, which remain rather flat between magic clusters, exhibiting a staircase profile.

Experiments on noble metal clusters ($Cu_N$, $Ag_N$, $Au_N$) indicate the existence of shell-effects, similar to those observed in alkali clusters. These are reflected in the mass spectrum [10] and in the variations of the ionization potential with $N$. The shell-closing numbers are the same as for alkali metals, that is $N = 2, 8, 20, 40$, etc. Cu, Ag and Au atoms have an electronic configuration of the type $nd^{10}(n + 1)s^1$, so the DFT jellium model explains the magic numbers if we assume that the $s$ electrons (one per atom) move within the self-consistent, spherically symmetric, effective jellium potential.

# 5 Electronic Shell Effects in Large Clusters

As the size of an alkali metal cluster increases, the gaps between electronic energy levels become smaller [12]. Eventually, when $N$ is sufficiently large, the discontinuous energy levels evolve into the quasicontinuous energy bands of the solid. When does this occurs? In other words, when are shell effects no longer discernible? Experiments indicate that shell effects remain important up to clusters with a few thousand valence electrons [13–16]. As an example, Table 1 lists the shell-closing numbers observed by Martin et al. [14] for sodium clusters with sizes up to $\sim 850$. These shell-closing numbers are revealed by large drops in the measured ionization potential.

The magic numbers appear at approximately equal intervals when the mass spectrum is plotted on a $N^{1/3}$ scale. More precisely, $\Delta N^{1/3} \approx 0.6$ between two consecutive magic numbers, where $N^{1/3}$ gives the linear dimension of the clusters [14]. One can understand qualitatively why shell structure should occur at approximately equal intervals on an $N^{1/3}$ scale [14]. Note that an expansion of $N$ in terms of the shell index K will always have a leading term proportional to $K^3$. One power of K arises because we must sum over all shells up to K in order to obtain the total number of particles. A second power of K arises because the number of subshells in a shell increases approximately linearly with shell index.

Finally, the third power of K arises because the number of particles in the largest subshell also increases with the shell index. Then:

$$N_K \sim K^3 .$$ (13)

When the number of electrons in the cluster increases, the number of electronic subshells also increases. Nevertheless, theoretical calculations have shown that groups of energy levels bunch together, leaving sizable empty gaps between them [17, 18]. However, do theoretical calculations for large clusters produce the precise bunching of energy levels that is required to explain the magic numbers observed in the experiments [13–16]? Although handling such a large number of electrons becomes more difficult, DFT calculations (a) lead to the bunching effect, that is to the $N^{1/3}$ periodicity, and (b) give magic numbers in close agreement with experiment.

The results of spherical jellium calculations performed by Genzken for Na clusters [18] are displayed in Fig. 3. After the cluster energy has been calculated as a function of $N$, it can be conveniently separated into a smooth and an oscillating part:

$$E(N) = E_{av}(N) + E_{shell}(N)$$ (14)

which is in accord with the idea of Strutinky's shell correction theorem [19]. The *liquid drop model* can be used to write $E_{av}(N)$ as the sum of a (negative) volume term, a surface term and a curvature term [20]:

$$E_{av}(N) = e_b N + a_s N^{2/3} + a_c N^{1/3} .$$ (15)

The bulk energy per atom, $e_b$, is obtained from the theory of the homogeneous electron gas [9]:

$$e_b = \frac{3}{10} (3\pi^2)^{2/3} (n^o)^{2/3} - \frac{3}{4} \left(\frac{3}{\pi}\right)^{1/3} (n^o)^{1/3} + e_c(n^o) .$$ (16)

As the electrostatic contributions cancel out, $e_b$ just contains kinetic (first term), exchange (second term) and correlation (third term) contributions and is a function of the average valence electron density $n^o$.

In addition the jellium model for a planar surface can be employed to calculate $a_s$ [21]. However, Genzken obtained $a_s$ from a plot of the slope of $(E(N)/N - e_b)$ versus $N^{-1/3}$, to suppress the shell oscillations for large values of $N$. Finally the curvature energy $a_c$ was fixed in a similar way by the slope of a plot of $E(N) - e_b N - a_s N^{2/3}$ versus $N^{-1/3}$ .

Subtracting the average part $E_{av}(N)$ from $E(N)$ defines the shell correction term $E_{shell}(N)$ in Eq. (14). This term is the energy actually plotted versus $N^{1/3}$ in Fig. 3. The pronounced oscillations exhibit sharp minima at the shell-closing numbers. The differences between these and the experimental magic numbers [15, 16] are rather small. The amplitude of the shell oscillations varies with size: the shell oscillations are enveloped by a slowly varying amplitude, the super-shell. The shell effects vanish periodically, but with a much larger size scale

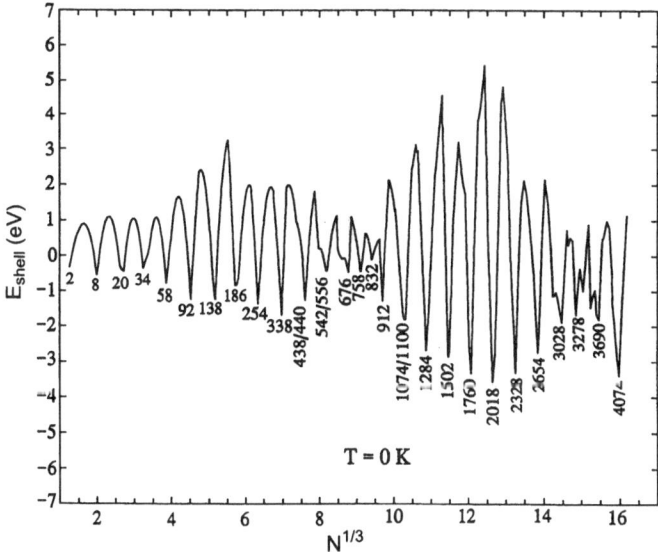

**Fig. 3.** The periodically varying contribution of valence electrons to the binding energy of a spherical sodium cluster ($E_{shell}$ in eq. (14)). Magic numbers are indicated. After Ref. [18].

($\Delta N^{1/3} = 6$). The first supershell node occurs at $N \approx 850$). Calculations by Nishioka et al. [17], using a nonselfconsistent Woods–Saxon potential (instead of the spherical jellium model) give $N \approx 1000$. This node has been observed, although the experiments also show some internal discrepancies: the first node is located at $N \sim 1000$ in Ref. [15] while it is at $N \approx 800$ in [16]. The experimental discovery of supershells confirms the predictions of nuclear physicists. However, supershells have not been observed in nuclei due to an insufficient number of particles. In summary, the existence of supershells is a rather general property of a system formed by a large number of identical fermions in a confining potential.

The supershell structure of lithium clusters has also been studied by Genzken [18]. The agreement with the experimental data [22] is even better than in the case of sodium. The experimental and the theoretical supershell node is found at $N \approx 820$.

The effect of finite temperature on the shells and supershells has been analyzed by Genzken for sodium clusters. For this purpose, calculations of the cluster "free energy" were performed by treating the valence electrons as a canonical ensemble in the heat bath of the ions [23]. (The spherical jellium model is even better at finite temperature.) Finite temperature leads to decreasing amplitudes of shell and supershell oscillations with increasing $T$. This is particularly important in the region of the first supershell node at $N \approx 850$, which is smeared out already at a quite moderate temperature of $T = 600$ K. However, temperature does not shift the positions of the magic numbers.

Similar shell and supershell effects have been observed in the trivalent metals Al, Ga and In [24–26]. However, in order to explain the details, it is necessary to go beyond the simple spherical jellium model. Although we present the improvements over the SJM below, we end this section by pointing out that the analysis of fine details suggests the need to go beyond the SJM even in the case of alkali metal clusters. An early example concerning large clusters is provided by the work of Lange et al. [27]. These authors have also performed DFT calculations of the ionization potential based on the spherical jellium model, with the objective to understand the drops in the measured ionization potentials of alkali metal clusters. However, to obtain the precise details in the drops of the ionization potential, the homogeneous jellium background had to be deformed slightly, by making the inner part of the clusters more dense. The results obtained by this inhomogeneous spherical jellium model are given in Table 1.

# 6 Impurities in Simple Metal Clusters

By supersonic expansion of mixed vapours, Kappes and coworkers [28] have obtained clusters containing a small amount of impurity atoms. In particular, we concentrate here on a series of clusters with the formula $A_N B$, that is, the cluster contains $N$ atoms of type A (alkali element) and a single impurity of type B (monovalent or divalent). The systems studied are listed in Table 2, along with the experimental abundance maxima in the small-size range. The order chosen for the list in the Table is that of increasing values of $\Delta n^o_+ = n^o_+(B) - n^o_+(A)$,

**Table 1.** Total number of electrons in closed shell sodium clusters [14]

| Shell | Experiment | Inhomogeneous Jellium |
|-------|------------|------------------------|
| A | 2 | 2 |
| B | 8 | 8 |
| C | 18 | 18 |
| D | 20 | 20 |
| E | 34 | 34 |
| F | 40 | 40 |
| G | 58 | 58 |
| H | 68 | 68 |
| I | 92 | 92 |
| J | 138 | 138 |
| K | $198 \pm 2$ | 196 (198) |
| L | $263 \pm 5$ | 268 |
| M | $341 \pm 5$ | 338 |
| N | $443 \pm 5$ | 440 |
| O | $557 \pm 5$ | 562 |
| P | $700 \pm 15$ | 704 |
| Q | $840 \pm 15$ | 854 |

**Table 2.** Abundance maxima observed in hetero-atomic $A_N B$ clusters. The abundance maxima are characterised by the number of valence electrons in the cluster. Also given is the difference $\Delta n^o_+$ between the jellium background densities of $A$ and $B$ metals

| $A/B$ | $n^o_+(B) - n^o_+(A)$ | Maxima |
|---|---|---|
| Na/Ba | 0.0008 | 8, 18 |
| Na/Sr | 0.0014 | 8, 18–20 |
| K/Na | 0.0018 | 8, 20 |
| Na/Eu | 0.0023 | 8, 18 |
| Na/Li | 0.0029 | 8, 20 |
| Na/Ca | 0.0029 | 8, 20 |
| Na/Yb | 0.0033 | 8, 20 |
| K/Li | 0.0047 | 8, 20 |
| Na/Mg | 0.0088 | 8–10 |
| K/Mg | 0.0106 | 10, 20 |
| K/Hg | 0.0106 | 10, 21 |
| Na/Zn | 0.0155 | 10, 20 |
| K/Zn | 0.0173 | 10, 20 |

where $n^o_+(B)$ and $n^o_+(A)$ are the jellium density parameters of the pure metals B and A respectively (see Eq. (5)).

The main feature in the Table is the observation of a new magic number, corresponding to ten valence electrons, for a large enough value of $\Delta n^o_+$. This new magic number begins with the system Na/Mg in this list. An associated feature that occurs earlier in the list is the vanishing of the magic number $N_e = 18$. Baladrón and Alonso [29] have demonstrated that the origin of the new magic number is "again" a shell-effect.

In general, the presence of the impurity atom induces a strong perturbation of the electronic cloud of an alkali cluster. The different nature of the impurity can be accounted for by a simple extension of the jellium model. The foreign atom is assumed to be at the cluster centre, and both subsystems – impurity and host – are characterized by different ionic densities in a jellium-like description. The following positive-charge background is then assumed:

$$n_+(r) = \begin{cases} n^o_+(B), & r < R_B \\ n^o_+(A), & R_B < r < R \\ 0, & r > R \end{cases} \tag{17}$$

$R_B$ is the Wigner–Seitz radius of metal $B$; that is, the radius of a sphere with volume $\Omega(B)$ and $R$ is the cluster radius, easily obtained from $\Omega(A)$, $\Omega(B)$ and the number of atoms.

All values of $n^o_+(B) - n^o_+(A)$ in Table 2 are positive, that is, the impurity provides a more attractive potential than the host. As a consequence, the original energy gap between the *2s* and *1d* levels is reduced. This is because the s-type electrons have a large probability of being near the center of the cluster, where the potential has become more attractive; enhancing the stability of the *2s* electrons. When the attractive power of the impurity is strong enough, the order of the *1d* and *2s* levels is reversed, leading to a new level ordering *1s, 1p, 2s, 1d,* that differs from that in the pure jellium model (*1s, 1p, 1d, 2s*). Ten electrons fill the first three subshells $(1s)^2 (1p)^6 (2s)^2$, so $N_e = 10$ appears as a magic number. Evidently, for this level ordering, the next magic number is $N_e = 20$, corresponding to the configuration $(1s)^2 (1p)^6 (2s)^2 (1d)^{10}$. In other words, the appearance of $N_e = 10$ is associated with the absence of $N_e = 18$. A quantitative view of this effect is given in Fig. 4, where we have plotted the stability function:

$$S(N_e) = E(A_{N-1}B) + E(A_{N+1}B) - 2E(A_N B) . \tag{18}$$

Here $S$ is written as a function of the number of valence electrons in the $A_N B$ cluster; this number is given by $N_e = NZ(A) + Z(B)$. Since $Z(A) = 1$, the clusters $A_{N-1}B$ and $A_{N+1}B$ differ from $A_N B$ by one electron. The cluster energies needed in Eq. (18) have been calculated by the same density functional technique described above for pure metal clusters. At the top of Table 2 we have Na/Ba. $\Delta n_+^0$ is very small in this case, and the order of the subshells is the usual one of

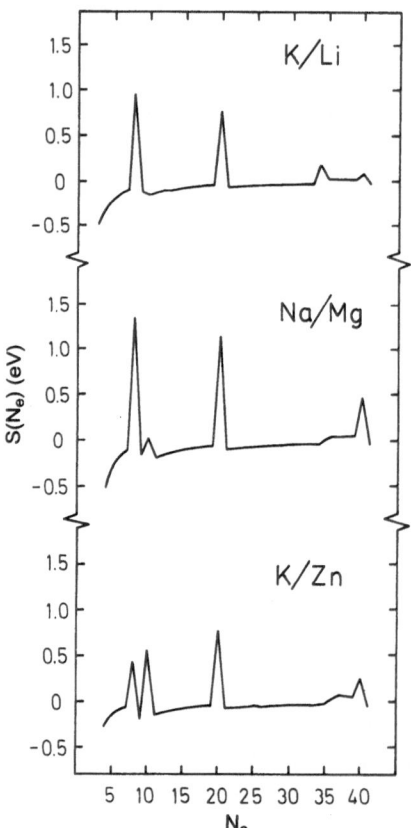

Fig. 4. Second derivative of the cluster energy as a function of the number of valence electrons in the cluster. Redrawn after data in Ref. [29].

the standard jellium model. As the attractive power of the impurity atom increases, the peak in $S(N_e)$ for $N_e = 18$ first decreases in magnitude and then disappears. This is so because the $2s$ level comes so close to the $1d$ level that the gap separating them vanishes. Finally, on increasing the attractive strength of the impurity even more, the $2s$ level becomes more stable than the $1d$ level, giving rise to the new shell closing numbers. The new peak at $N_e = 10$ appears at Na/Mg, just as in the experiment.

# 7 The Spheroidal Model

The spherical assumption is very successful in explaining the most prominent features of the ionization potential and the mass spectra of simple metal clusters. However, there is evidence of many small features which the SJM is unable to explain. Whenever a top-shell is not completely filled $(N \neq 2, 8, 20, ...)$, the electronic density becomes non-spherical, which, in turn, leads to a distortion of the ionic background. This Jahn-Teller type distortion, similar to those observed for molecules and nuclei, leads to a splitting of all spherical shells into sub-shells [30]. Deformed clusters are prevalent for open-shell configurations. Clemenger [31] has studied the effect of deformations for alkali clusters, using a modified-three-dimensional harmonic oscillator potential. The model considers different oscillation frequencies along the $z$ axis, (chosen as the symmetry axis; and perpendicular to it. The model Hamiltonian used by Clemenger also contains an unharmonic term, that serves to flatten the bottom of the potential well. Due to the deformation, the highly degenerate spherical shells are split into sub-shells.

The jellium background model has been extended by Ekardt and Penzar to account for spheroidal deformations [30, 32]: the ionic background is represented by a distribution of positive charge with constant density and a distorted, spheroidal shape. The advantage over Clemenger's model is that the spheroidal jellium model is parameter-free and that the calculation of the electronic wave functions is performed self-consistently. Due to the cylindrical symmetry of the problem, the azimutal quantum number $m$ is still a good quantum number for the electronic states, as is the parity with respect to the reflection at the midplane. However, the angular momentum $l$ is no longer a good quantum number and, as a consequence, the problem is intrinsically two-dimensional. This makes the Kohn–Sham equations harder to integrate.

Assuming major axes $a$ and $b$ for an axially symmetric spheroid, a distortion parameter $\eta$ can be defined:

$$\eta = \frac{2(a - b)}{a + b} .$$ (19)

The deformation parameter $\eta$ describes how prolate or oblate the cluster is. This distortion parameter is determined for each cluster by minimizing its total energy. The main first-order effects of the spheroidal model are energy splittings proportional to $\eta$. These lead to fine structure in the stability function $\Delta_2(N)$ (see Eq. (12)), which has, in addition to the usual peaks of the spherical jellium model, smaller subshell-filling peaks at $N = 10, 14, 18, 26, 30, 34, 36, 38, 44, 50, 54$, etc. All the fine-structure peaks predicted by $\Delta_2(N)$ are observed in the experimental mass spectra [6]. Some examples of the agreement follow: the fourfold patterns in the $1f$ and $1g$ shells appear correctly, as well as the twofold pattern in the $2p$ shell, corresponding to the filling of a prolate subshell at $N = 36$ and an oblate shell at $N = 38$.

# 8 General Discussion of the Ionization Potential

The ionization potential IP is the energy necessary to extract one electron from the neutral cluster. For a macroscopic solid this is called the work function, $W$. In this case Lang and Kohn [33] have shown that $W$ can be expressed as the sum of three terms

$$W = D_{es} + \mu_{xc} - E_F \tag{20}$$

where all contributions are taken to be positive. The first, electrostatic term, $D_{es}$, represents the surface dipole barrier, resulting from the spilling of electronic charge beyond the positive jellium background boundary. The second, $\mu_{xc}$, is the exchange and correlation contribution to the chemical potential of an uniform electron gas. These two terms mainly determine the depth of the potential well. The kinetic energy term, $E_F$, is the bulk Fermi energy. Perturbative inclusion of the ion pseudopotentials decreases the calculated values by $\approx 10\%$ and leads to work functions for the alkali metals in reasonable accord with experiment Similar agreement was also obtained for other simple metals.

When the size of the metallic piece is microscopic, a correction term is required. If this correction is calculated from simple classical electrostatic considerations as the energy required to remove an electron from a metallic sphere of radius $R$, the following result is obtained for the ionization potential [34]:

$$IP = W + \frac{1}{2}\frac{e^2}{R} \tag{21}$$

and the corresponding expression for the electron affinity $EA$ is:

$$EA = W - \frac{1}{2}\frac{e^2}{R} . \tag{22}$$

Theoretical considerations based on density functional theory, which transcend the simple electrostatic arguments, indicate that a more correct form of these equations is [35]:

$$IP = W + \left(\frac{1}{2} - c\right)\frac{e^2}{R} \tag{23}$$

$$EA = W - \left(\frac{1}{2} + c\right)\frac{e^2}{R}. \tag{24}$$

The constant $\frac{1}{2}$ comes from the classical electrostatic energy and $c$ is a material-dependent constant, arising from the electronic kinetic and exchange-correlation energies. These equations are valid for $R$ that is large compared to the atomic radius, that is, when shell effects become negligible. The exact value of $c$ is not known. DFT calculations for the SJM [20, 36] give $c \approx 0.08$ or a little bit larger ($c \approx 0.14$) [37]. These values, in particular the last, are consistent with the empirical value $c \approx 0.12 \pm 0.06$ obtained from a photoemission study of very large Ag clusters containing 5000–40000 atoms [38]. Numerous studies of medium size clusters, $N \leqslant 100$, of different metallic elements give a good fit to experiment with $c = 1/8 = 0.125$ [39, 40].

The general agreement of spherical droplet predictions with the ionization potential and electron affinity data has several implications: (1) The assumption of spherical symmetry is viable ($N > 10$). (2) The size dependence of IP and EA is overridingly determined by changes in curvature above the level of quantum size effects, which are typically no larger than 10% of the IP. (3) Valence electrons are delocalized, even for very small clusters.

## 9 Odd–Even Effects in the Ionization Potentials

Superimposed on the smooth behavior described by equations (23) and (24), the experimental data on the ionization potentials and electron affinities show two additional features. One, which has already been discussed, is the shell-closing effect. The second effect, which can be observed in clusters of monovalent $s$-electron metals, is an odd–even effect, also apparent in the mass spectra. Some examples follow:

(i) In ionization potential measurements of alkalis ($Na_N, K_N, N < 20$), $N$-even clusters systematically have slightly larger values (by 0.1–0.2 eV) than their $N$-odd neighbors [41].
(ii) An inverse effect is found for the electron affinity of noble metal clusters, with $N$-odd clusters having higher photodetachment thresholds [42].
(iii) The mass spectra of both positive and negative noble-metal clusters obtained by ion bombardment, shows an odd–even alternation in the abundances, with $N$-even clusters being less abundant than their $N$-odd neighbors [10, 43]. This effect is observed up to $N \approx 40$.

*Ab initio* DFT calculations, using the local-spin-density approximation for exchange and correlation, reproduce the odd–even effects in the ionization potentials and binding energies [44]. In these calculations, the granular structure of the ionic cores is retained, although these are often replaced by (non local) pseudopotentials. The calculations are difficult, because the geometrical conformation of the cluster (that is, the positions of the ions) has to be calculated by minimizing the total energy.

The odd–even effect results from the interplay between cluster deformation and spin effects. In Fig. 5 we show the evolution of the *molecular orbitals* for the calculated most stable geometrical conformations of alkaline clusters with sizes $N \leqslant 14$. First of all, there is a smooth increase of the binding energy of the *1s* orbital with increasing cluster size. (We use a notation which reflects the nodal character of the molecular orbitals and allows relating them to the orbitals of the jellium model.) Also the binding of the manifold of *1p*-type levels shows an overall increase with increasing $N$. However, contrary to the predictions of the spherical jellium model, the $1p_x$, $1p_y$ and $1p_z$ orbitals are not degenerate. The splitting occurs because the cluster, and thus the effective DFT potential acting

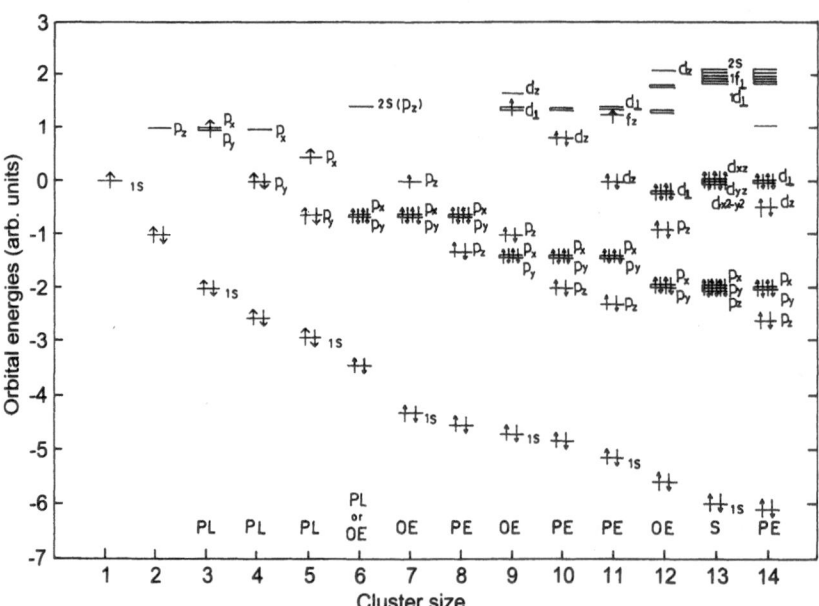

**Fig. 5.** Evolution of the molecular orbitals of alkali metal clusters with cluster size, corresponding to the most stable structure for each cluster size [45]. Orbital energies are in arbitrary units. PL and S denote planar and spherical structures, OE and PE pertain to oblate and prolate ellipsoids, respectively. Redrawn from data in Ref. [45].

on the electrons, is not spherical. The characteristic shape of the cluster is indicated at the bottom of the picture. The magnitude of the energy difference between $(1p_x, 1p_y)$ and $1p_z$ orbitals reflects the degree of distortion from spherical symmetry. The $1p_z$ orbital has a lower binding energy than $(1p_x, 1p_y)$ in oblate clusters and the order is reversed for prolate ones. When this splitting is combined with the fact that double occupation of a $p$-orbital increases its binding energy over that of single occupation (spin-pairing effect), then the odd–even effect in both the ionization potential and relative cluster stability (mass abundance) is explained. The splitting of the $p$ levels is a self-consistent effect. When the $p$-shell is not fully occupied, the electron density is not spherically symmetric. This, in turn, induces a distortion of the cluster geometry away from the spherical shape, that leads to a splitting of the $p$-levels. Similar arguments concerning the splitting of the $d$-shell, etc, rationalize the odd–even effect for larger clusters. In summary, the origin of the odd–even effect is a Jahn-Teller-type deformation of the ground state of the cluster from its spherical shape, leaving only the double degeneracy of each level due to spin.

The strong fluctuation of IP or of the mass abundance is an electronic-structure effect, reflecting the global shape of the cluster, but not necessarily its detailed ionic structure. This is demonstrated in Fig. 6, where the ionization potentials of sodium clusters obtained by the spheroidal jellium model [32] are compared with their experimental values [46]. The odd–even oscillation of IP for low $N$ is reproduced well. The amplitude of these oscillations is exaggerated, but this is corrected by using the spin-dependent LSDA, instead of the simple LDA [47]. The same occurs for the staggering of $\Delta_2(N)$ [48].

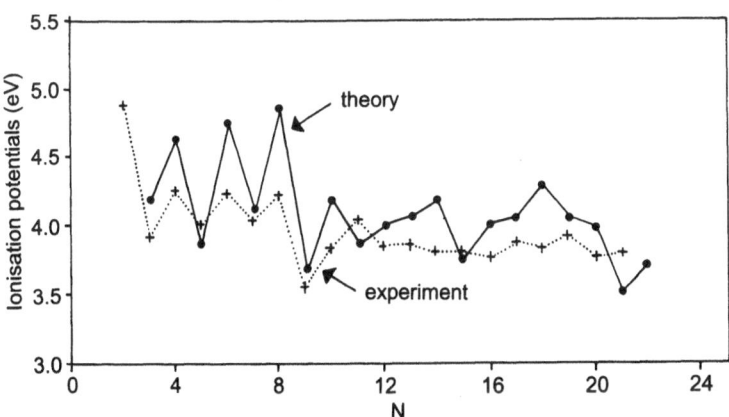

**Fig. 6.** Comparison of the ionization potential of sodium clusters, obtained with the spheroidal jellium model [32] and experiment [46]. Redrawn from data in Ref. [32].

## 10 Response to a Static Electric Field

The net force acting on a neutral system in the presence of an static (time independent) inhomogeneous electric field $E(r)$ can be expanded as:

$$F = E(r_0) \int \Delta n(r) \, dr + \nabla E(r_0) \int \Delta n(r)(r - r_0) \, dr$$

$$+ \frac{1}{2} \nabla^2 E(r_0) \int \Delta n(r)(r - r_0)^2 \, dr + \ldots \tag{25}$$

where $\Delta n(r) = n(r) - n_+(r)$, and $n(r)$ and $n_+(r)$ are the electronic and ionic densities, respectively. The first term does not contribute in a neutral system, and the second term is the product of the field gradient $\nabla E(r_0)$ times the induced dipole moment $p$ of the system. If the deformation of the ionic density due to the applied electric field is small, then the force on the system is, in first order

$$F \approx p \nabla E(r_o) \tag{26}$$

where $p$ can be written as the product $p = \alpha E(r_0)$ of the polarizability $\alpha$ and the applied field. Based on these ideas, Knight and coworkers determined the polarizability of neutral alkali clusters by measuring the deviation of a cluster beam that travels through a region where an inhomogeneous electric field has been applied [49]. The polarizabilities of Al clusters were also determined by the same method [50].

Within DFT, the polarizability, $\alpha$, of a cluster can be calculated using linear response theory [51]. We begin by considering the ground state of the system. The ground-state density is calculated, as usual, by solving the single-particle Kohn–Sham equations. Let us now apply a static electric field, characterized by a multipole potential of the form $\delta V_L = E_o r^L Y_L^o(\theta)$, where $E_o$ is a small number and $Y_L^o(\theta)$ is a spherical harmonic. The system then develops an induced moment $p_L$ of magnitude $p_L = \alpha_L E_o$ in response to the field. The first-order response of the system is characterized by a small change in the one-electron wave functions $\phi_i(r) \to \phi_i(r) + \delta\phi_i(r)$. The corresponding change in the electron density can be written,

$$n(r) = \sum_{i=1}^{occ} |\phi_i(r)|^2 \to n(r) + \delta n(r) \tag{27}$$

where $\delta n(r)$ is given by $\delta n(r) = 2Re\left[\sum_{i=1}^{occ} \phi_i^*(r)\delta\phi_i(r)\right]$. Using first-order perturbation theory and the Kohn–Sham equations, one obtains a set of equations for the changes $\delta\phi_i$ in the wave functions when the perturbing field is present:

$$\left[\frac{1}{2}\nabla^2 + V_{eff} - \varepsilon_i\right]\delta\phi_i(r) = \delta V_{eff}(r)\phi_i(r) \tag{28}$$

where:

$$\delta V_{eff}(r) = E_o r^L Y_L^o(\theta) + \int d^3 r' \frac{\delta n(r')}{|r - r'|} + \delta n(r) \frac{\partial V_{xc}^{LDA}}{\partial n(r)} \tag{29}$$

is the self-consistent potential associated with the change in electron density due to the external field.

The calculation of the polarizability, $\alpha_L$ proceeds by first solving the Kohn–Sham equations for the field-free cluster, and thus obtaining $\phi_i$ and $\varepsilon_i$. Then, after solving Eqs. (28) and (29) self-consistently to get the set of functions $\delta\phi_i(r)$, $\delta n(r)$ is evaluated, and the $L$-order polarizability is finally obtained as:

$$\alpha_L = (1/E_o) \int d^3 r \, r^L Y_L^o(\theta) \, \delta n(r) \ . \tag{30}$$

In the case of $L = 1$ (dipole polarizability) this formula reduces to:

$$\alpha_{L=1} = \frac{1}{E_o} \int z \delta n(r) \, d^3 r \ . \tag{31}$$

The calculated electric dipole polarizabilities of some Na clusters with closed electronic shells are given in Table 3 (see column LDA–SJM) [52]. The calculations employed the spherical jellium model. The results are expressed in units of the classical polarizability $R^3$. The enhancement of $\alpha$ over its classical value is directly proportional to the fraction of the electronic charge that extends beyond the positive background in the field-free system. The agreement with experiment is reasonable, although the theory systematically underestimates the polarizabilities.

To improve the agreement with experiment, two kinds of corrections have been applied. The first consists in smoothing the discontinuity of the jellium density at the cluster surface. For this purpose, the original step-density is replaced by a continuous function that models a surface with a finite thickness of about 1 a.u. [52]. As a consequence, the electron density is more extended and the polarizability increases in this finite surface jellium model (see column FSJM), improving the agreement with experiment. The FSJM also improves

**Table 3.** Electric dipole polarizabilities in units of $R^3$, of neutral sodium clusters in the spherical jellium model (SJM) and in a jellium model with finite surface thickness (FSJM)

| N | LDA [52] | | Exp [49] | WDA [54] |
|---|---|---|---|---|
| | SJM | FSJM | | |
| 8 | 1.45 | 1.71 | 1.77 ± 0.03 | 1.81 |
| 18 | 1.33 | 1.53 | | |
| 20 | 1.37 | 1.61 | 1.68 ± 0.10 | 1.63 |
| 34 | 1.27 | 1.46 | | |
| 40 | 1.32 | 1.56 | 1.61 ± 0.03 | 1.53 |

other fundamental properties. For instance, Fig. 2 shows that the magic character of $N = 40$ is weaker than that of $N = 34$ in the SJM, which is contrary to experiment. The FSJM increases the $2p - 1g$ energy gap in the single-particle spectrum, and thus increases the magic character of $N = 40$, restoring agreement with experiment [52].

Other correction consists in replacing the LDA by a more accurate, nonlocal description of exchange and correlation. In a neutral cluster, $V_{xc}^{LDA}(r)$ goes to zero exponentially at larger $r$. However, in the non-local weighted-density approximation (WDA) [53], the asymptotic behavior of $V_{xc}$ is proportional to $(-1/r)$. This slower decay gives a more extended density tail, and consequently a higher polarizability [54]. This is shown in the column WDA of Table 3. These calculations also employed the spherical jellium model.

# 11 Dynamical Response

Using a method analogous to that for static case, the linear response theory can be developed within the LDA for the case when the external electric field, characterized by the potential $V_{ext}(r; \omega) = E_o r^L Y_L^o e^{i\omega t}$, is time-dependent. This leads to the *time-dependent density functional theory* (TDLDA) [55].

In this case, the external field induces a time-dependent perturbation of the electron density of the cluster, $\delta n(r, t)$, with Fourier components $\delta n(r, \omega)$. The key quantity for calculating the response of the system in the linear response regime is the dynamic susceptibility $\chi(r, r'; \omega)$, which relates the individual components, $\delta n(r, \omega)$, of the induced density to those of the applied field:

$$\delta n(r; \omega) = \int \chi(r, r'; \omega) V_{ext}(r'; \omega) \, dr'. \tag{32}$$

Again, the interest is primarily on the case of a dipole field, for which $V_{ext}(r, \omega) = E_0 z e^{i\omega t}$. The dynamical polarizability $\alpha(\omega)$, which, in this case $(L = 1)$ is the ratio of the induced dipole moment and the external field strength, becomes:

$$\alpha(\omega) = \frac{1}{E_o} \int z \delta n(r; \omega) \, dr \tag{33}$$

The dynamical polarizability evidently reduces to the static one of Eq. (31) in the case of $\omega = 0$. Using *Fermi's golden rule*, one finally obtains the photoabsorption cross section of the cluster,

$$\sigma(\omega) = \frac{4\pi\omega}{c} \, \text{Im} \, \alpha(\omega) \tag{34}$$

where $c$ is the velocity of light and $\text{Im} \, \alpha(\omega)$ indicates the imaginary part of $\alpha(\omega)$.

Let us now return to the calculation of $\chi(r, r'; \omega)$. In its Kohn–Sham formulation, DFT is a theory of independent particles moving in an effective self-consistent field. Thus Eq. (32) can be rewritten,

$$\delta n(r; \omega) = \int \chi_0(r, r'; \omega) \delta V_{eff}(r'; \omega) d^3 r' \tag{35}$$

where $\chi_o(r, r'; \omega)$ is the independent-particle (or noninteracting) susceptibility, and $\delta V_{eff}(r'; \omega)$ is the self-consistent perturbing potential (cf. Eq. (29)):

$$\delta V_{eff}(r'; \omega) = V_{ext}(r'; \omega) + \int \frac{\delta n(r''; \omega)}{|r' - r''|} d^3 r'' + \int \frac{\delta^2 E_{xc}}{\delta n(r') \delta n(r'')} \delta n(r''; \omega) d^3 r''. \tag{36}$$

A convenient way of writing $\delta V_{eff}$ is

$$\delta V_{eff}(r'; \omega) = V_{ext}(r'; \omega) + \int K(r', r'') \delta n(r''; \omega) d^3 r'' \tag{37}$$

where we have introduced the nonlocal Kernel, or residual particle-hole interaction:

$$K(r', r'') = \frac{1}{|r' - r''|} + \frac{\delta^2 E_{xc}}{\delta n(r') \delta n(r'')}. \tag{38}$$

If the LDA is used for $E_{xc}$, the local field correction, which is the last term in (38), becomes a local function.

Inserting the expression (37) for $\delta V_{eff}$ in (35), and using the form (32) for $\delta n(r; \omega)$ on both sides of the resulting equation, the following Dyson-type equation is obtained for the interacting dynamic susceptibility:

$$\chi(r, r'; \omega) = \chi_0(r, r'; \omega) + \int \chi_0(r, r_1; \omega) K(r_1, r_2) \chi(r_2, r'; \omega) d^3 r_1 d^3 r_2 \tag{39}$$

which has to be solved iteratively.

Finally, the single-particle (or non-interacting) susceptibility, which is needed to solve Eq. (39) has the form

$$\chi_0(r, r'; \omega) = \sum_{i=1}^{occ} \{\phi_i^*(r) \phi_i(r') G(r, r'; \varepsilon_i + \hbar\omega)$$

$$+ \phi_i(r) \phi_i^*(r') G^*(r, r'; \varepsilon_i - \hbar\omega)\} \tag{40}$$

where the $\phi_i$ are the occupied single particle states of the ground state KS calculation for the field-free system, $\varepsilon_i$ are the corresponding single-particle energies and $G(r, r'; \varepsilon_i \pm \hbar\omega)$ are the retarded Green functions associated to the effective Kohn–Sham potential:

$$\left[ E + \frac{1}{2} \nabla^2 - V_{eff}(r) \right] G(r, r'; E) = \delta(r - r') \tag{41}$$

141

Before discussing the results of the theoretical calculations, let us explain the way experimentalists obtain the photoabsorption spectrum of metallic clusters. Knight's group was the first to measure photoabsorption spectra, that are based on the fact that the clusters fragment upon absorption of light of the appropriate frequency; this effect induces a deviation of the cluster away from the initial direction of the molecular beam [56]. The ratio between the number of clusters of a given size arriving at the detector with and without light excitation gives the value of the absorption cross section. The process involved in this depletion experiment is the excitation of a collective mode (surface plasmon) at an energy that is about 3 eV for sodium clusters. Since this energy is higher than the binding energy of an atom in the Na aggregate (approximately 1.1 eV), the excited cluster decays by evaporating single atoms. Using a statistical model and assuming that the energy of the collective mode is converted into vibrations, the time required to evaporate an atom turns out to be $10^{-12}$–$10^5$ seconds for clusters with a size between 8 and 40 atoms. This time is very short compared to the time of flight of the molecular beam in the spectrometer ($\approx 10^{-3}$ seconds). Consequently one can assume that the photoabsorption and photoevaporation cross sections are equal. The collective excitations (or surface plasmons) in metallic clusters are similar to the giant dipole resonances in nuclei.

The integral of $\sigma(\omega)$ gives rise to the dipole sum rule,

$$\int_0^\infty \sigma(\omega)\, d\omega = \frac{4\pi^2}{\hbar c} m_1 \tag{42}$$

where $m_1 = e^2 \hbar^2 Z / 2m$. $Z$ is the number of electrons taking part in the collective motion. Consequently, the experimental determination of $\sigma(\omega)$ helps identify the collective nature of a resonance.

The classical theory of dynamic polarizability developed by Mie predicts a single dipole resonance at a frequency given by

$$\omega_{Mie} = \sqrt{\frac{Z\hbar^3 c^2}{mR^3}} \tag{43}$$

where $m$ is the electron mass and $R = r_S Z^{1/3}$ is the cluster radius ($r_S$ is the radius of a sphere containing one electron.) This gives for $\omega_{Mie}$ a value equal to one third of the bulk plasma frequency, $\omega_{pl}$.

*Linear response theory* (TDLDA) applied to the jellium model follows the Mie result, but only in a qualitative way: the dipole absorption cross sections of spherical alkali clusters usually exhibit a dominant peak, which exausts some 75–90% of the dipole sum rule and is red-shifted by 10–20% with respect to the Mie formula (see Fig. 7). The centroid of the strength distribution tends towards the Mie resonance in the limit of a macroscopic metal sphere. Its red-shift in finite clusters is a quantum mechanical finite-size effect, which is closely related to the spill-out of the electrons beyond the jellium edge. Some 10–25% of the

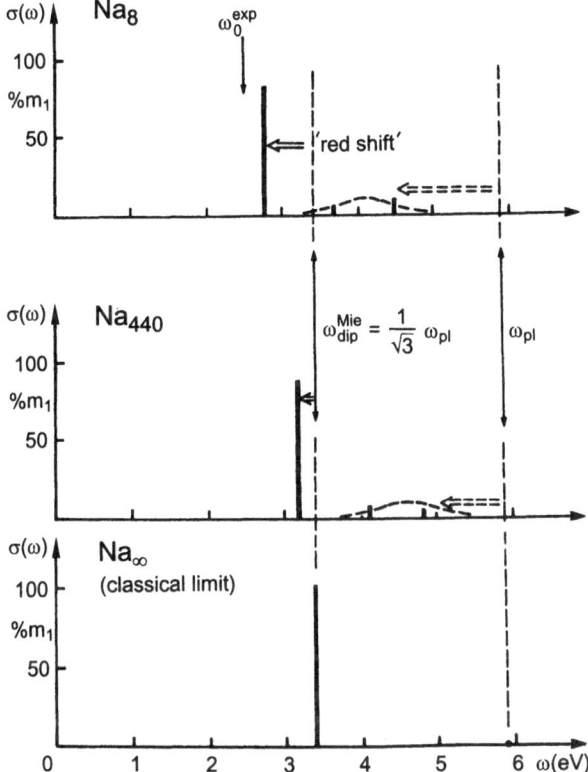

**Fig. 7.** Schematic representation of the collective dipole spectra of sodium clusters obtained in linear response theory [57]. The quantity plotted is $\sigma(\omega)$ as the percentage of the total dipole strength, $m_1$, normalized to 100% (see Eq. (42)). The lowest spectrum ($Na_\infty$) represents the classical limit, where 100% of the strength lies in the surface plasmon (frequency $\omega_{Mie}$) and the volume plasmon (frequency $\omega_{pl}$) has zero strength. For finite clusters the surface plasmon is red-shifted and its missing strength is distributed over the remainder of the strongly fragmented volume plasmon.

dipole strength is typically found at higher energies and can be interpreted as a reminiscence of a strongly fragmented volume plasmon.

Often the dominant peak is fragmented into two or more lines. The fragmentation of the collective strength in spherical clusters can be attributed to an interference of specific particle-hole (or more complicated) excitations with the predominant collective mode. This fragmentation may be compared to Landau damping in the solid, although there it refers to a collective state lying in a single-particle continuum.

As compared to experiment, all spherical jellium calculations yield an insufficient redshift of the Mie resonance. This is connected to the low polarizability. Therefore, a jellium density with a smooth surface [52], or other corrections found to improve the polarizability, also improve the position of the dipole resonance. Replacing the LDA by a nonlocal description of exchange and correlation within the time-dependent DFT provides a step in the right direction. Self-interaction corrections (SIC) have been applied by Pacheco and Ekardt [58] and the *weighted density approximation* (WDA) by Rubio et al. [59]. The behavior of the exchange-correlation potential is improved in the asymptotic (large r) region and the local field correction (see Eq. (38)) is also improved. The number of bound states increases as compared with LDA, and thus the Landau damping as well. The plasmon resonances are displaced toward lower frequencies, leading to better agreement with experiment. Nonlocal effects play an essential role when the plasmon excitation occurs near the ionization threshold. In this case fragmentation of the plasmon resonance is expected. This should be the case for negatively charged clusters of a certain size. For small negative clusters, the collective resonance frequency lies in the region of transitions to the continuum of states, where Landau damping produces a large broadening of the resonance. As the cluster size increases, the plasmon frequency approaches the region of discrete states, because its value changes at a much lower rate than the increase in the ionization threshold of the negative cluster. Then, when the plasmon frequency is close to this ionization threshold, fragmentation of the plasmon is expected. This expectation has been confirmed by calculations for large negatively charged potassium [59] and sodium clusters [60]. Fragmentation is obtained using the WDA, but not if the LDA is used instead. This is because the LDA predicts a lower ionization threshold (identified with $-\varepsilon_{HOMO}$, where HOMO indicates the lowest bound orbital of the "negative" cluster), which remains well separated from the plasmon resonance.

There are other deviations from the single-resonance Mie formula that are reproduced by the jellium model calculations [61]. In open-shell clusters a further splitting of the dipole resonance is observed; it is a consequence of their static deformation and can easily be described by the TDLDA calculations within the context of the spheroidal jellium model [61]. The double-peak feature in the photoabsorption cross section of positively charged clusters has been observed for $K_{11}^{+}$ by Brechignac [62], for Ag clusters in the region $10 \leqslant N \leqslant 16$ by Tiggesbäumker et al [63], and also for Na clusters [64]. The double peak indicates the two modes corresponding to excitation along the main axis of the spheroid and perpendicular to it. The results of Borggreen [64] reveal the systematics of cluster shapes observed when adding electrons to the $N = 8$ and $N = 20$ spherical clusters: spherical → prolate → oblate → spherical. This sequence is reproduced by the spheroidal jellium model [30, 32]. In larger clusters it is not easy to disentangle the effects of static deformations from those of the fragmentation mechanism discussed above.

The observed widths of the resonance peaks are more difficult to explain microscopically than their positions. The decay mechanisms of the collective dipole resonances are, both theoretically and experimentally, still rather poorly understood. More experimental information on their temperature dependence and on the detailed line form is required in order to shed light on this problem.

# 12 Triaxial Deformations

A splitting of the dipole resonance into three peaks has been observed in some sodium clusters [64]. This observation is interpreted as corresponding to collective vibrations of the valence electrons in the directions of the principal axes of a triaxially deformed cluster, and has motivated an extension of the deformed jellium model to fully triaxial shapes. Lauritsch et al. [65] have applied this model to $Na_{12}$ and $Na_{14}$.

The potential energy surfaces of these two clusters were calculated with the intention to study the splitting of the dipole resonance, as well as the competition between possible shape isomers. The triaxial (or ellipsoidal) deformations of the jellium density can be classified in terms of the Hill–Wheeler coordinates $(\beta, \gamma)$ for quadrupole deformations [66]. $\beta$ describes the overall quadrupole deformation. $\gamma = 0°, 120°, 240°$ describe prolate deformations and $\gamma = 60°, 180°, 300°$ describe oblate ones; all other values of $\gamma$ refer to truly triaxial shapes. In addition to the shape deformation of the positive background, Lauritsch et al. also allowed the jellium density to have a diffuse surface profile that could be modelled with a Fermi function. The ground state of $Na_{12}$ is predicted to be triaxial, with deformation parameters $\beta = 0.54$, $\gamma = 15°$, and is energetically well separated from competing prolate and oblate configurations. $Na_{14}$ is characterised by axially symmetric minima: the two lowest configurations, prolate and oblate respectively, are almost degenerate in energy. The oblate minimum is rather soft in the $\gamma$-direction whereas the prolate minimum predicts stiffer $\gamma$-vibrations. The pronounced shape isomerism found for both clusters bears some resemblance to that found by fully microscopic quantum chemical [67] or ab initio DFT calculations [68].

Lauritsch et al. [65] obtained the resonance energies of the collective dipole oscillations of the valence electrons from the approximation:

$$(\hbar\omega_i)^2 = \frac{\hbar^2}{Nm} \int d^3 r n(r) \frac{\partial^2}{\partial r_i^2} V_{ext}(r) \tag{44}$$

where $V_{ext}(r)$ is the electrostatic jellium potential, $m$ is the electron mass, and $i$ runs over the spatial directions, i.e. $r_i = \{x, y, z\}$ for triaxial clusters and $r_i = \{r, z\}$ for axial ones. Expression (44) is obtained from the random-phase approximation (RPA) sum rules [57, 69]. The resonance energies were calculated for the most prominent minima in both clusters and the results are shown in Table 4. Three different energies are obtained for the ground state of $Na_{12}$, reflecting its triaxial shape. The three energies are in qualitative agreement with the three experimental peaks [64], although the calculated energies are 10–15% too high due to the simple sum-rule approximation. Each of the two competing axial states of $Na_{14}$ is characterized by a double-peak structure where $\hbar\omega_r$ has double weight compared to $\hbar\omega_z$. The actual strength distribution will be an incoherent superposition of the two isomeric minima, but the precise

**Table 4.** Dipole surface plasmon energies in a.u. obtained by Lauritsch et al. [65] from the RPA dipole sum rule approximation (44).

| $Na_{12}$ | | | $Na_{14}$ | | | |
|---|---|---|---|---|---|---|
| triaxial | | | oblate | | prolate | |
| $\hbar\omega_x$ | $\hbar\omega_y$ | $\hbar\omega_z$ | $\hbar\omega_r$ | $\hbar\omega_z$ | $\hbar\omega_r$ | $\hbar\omega_z$ |
| 0.105 | 0.119 | 0.085 | 0.093 | 0.125 | 0.114 | 0.085 |

outcome is difficult to predict because it can depend sensitively on the formation process of the clusters.

Kohl et al. [47] have extended the calculations to a larger set of Na clusters (Na = 2–20). They confirm the results of the spheroidal jellium model: prolate clusters after the magic numbers $N = 2$ and $N = 8$ and oblate ones before $N = 8$ and $N = 20$. However, a transition region formed by triaxial shapes was found separating the prolate and oblate regimes. The width of this transition region is very small between $N = 2$ and $N = 8$, containing only the cluster $Na_5$, but comparatively large between $N = 8$ and $N = 20$. The triaxial minimum is well developed in $Na_5$ but triaxiality of the others is extremely soft so that thermal fluctuations easily wash out the triaxial signatures in the dipole resonance energies.

# 13 Fission of Charged Clusters

Stable multiply charged clusters can be observed only above a critical size, $N_c$, that depends on the metal and on the charge state. In the case of alkali metals, the critical size for the observation of doubly charged clusters is 25 for Li [70], 27 for Na [39, 71], 20 for K [70], and 19 for Cs [72]. Critical sizes of clusters as highly ionized as $Na_N^{7+}$ and $Cs_N^{7+}$ have also been determined [73].

These critical sizes are determined studying charged clusters produced by multi-step ionization of hot larger clusters that lose a sizeable part of their excitation energy by evaporating neutral atoms. This causes the ionized clusters $X_N^{q+}$ to shrink up to sizes $N \approx N_c$. For sizes around $N_c$, monomer evaporation competes with another dissociation channel, asymmetric fission, in which two charged fragments of different size are emitted. Electronic shell effects are manifested in the fission channel, with preferential emission of closed-shell fragments, such as $Na_3^+$, $K_3^+$, or $K_9^+$ [74].

However, stable multiply charged clusters have also been detected below $N_c$. This occurs if they are created in a multi-step ionization process starting from cold neutral clusters [39, 71], and indicates the existence of a stabilizing fission barrier. These experimental facts show that cluster fission is a barrier-controlled process, and have prompted interest in calculating these barriers. Old DFT studies of charged cluster fragmentation using the SJM were based on purely-energetic criteria, which only involve the energies of the initial and final states [75]. These early studies correctly predicted that emission of closed shell fragments ($Na_3^+$, $Na_9^+$, ...) is likely to occur.

However, consideration of the fission barrier is necessary in order to understand the nature of $N_c$ and calculate its value. For hot large clusters the preferred decay channel is evaporation,

$$X_N^{2+} \rightarrow X_{N-1}^{2+} + X \tag{45}$$

because the barrier against fission is larger than the heat $\Delta H_e$ required to evaporate a neutral atom. On the other hand, small clusters undergo fission:

$$X_N^{2+} \rightarrow X_{N-3}^{2+} + X_3^+ \tag{46}$$

because the fission barrier is, in this case, smaller than $\Delta H_e$. (In (46) we have assumed, for simplicity, that the most favourable fission channel is the emission of a charged trimer.)

A schematic representation of these two cases is shown in Fig. 8. $\Delta H_e$ is given by the energy difference:

$$\Delta H_e = E(X_{N-1}^{2+}) + E(X) - E(X_N^{2+}) \tag{47}$$

and $F_m$, the fission barrier height, is the difference in energy between the fissioning cluster at the saddle point and the parent cluster:

$$F_m = E(saddle) - E(X_N^{2+}) . \tag{48}$$

If the fission barrier vanishes, the cluster is absolutely unstable with respect to fission.

So far, no experimental *model-independent* determinations of fission barrier heights are available. A promising method has been proposed by Bréchignac et al. [70] which employs two ingredients: one is the experimental branching ratio of fission to monomer evaporation corresponding to the same doubly

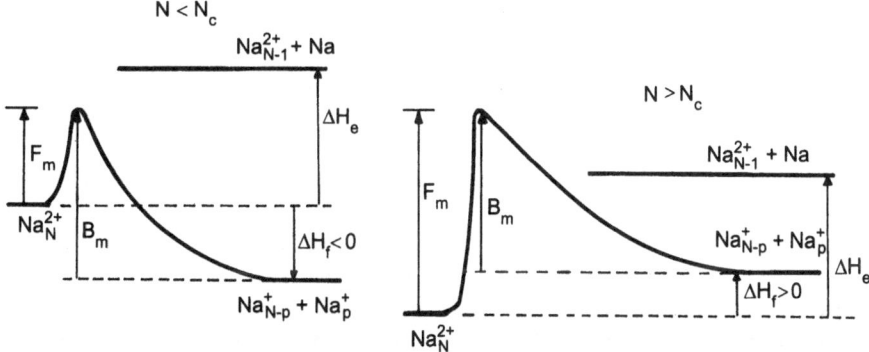

**Fig. 8.** Competition between fission and evaporation. $\Delta H_e$ and $\Delta H_f$ are the heats of evaporation and fission, respectively. $F_m$ and $B_m$ are the fission and fusion barrier heights, respectively.

charged cluster, and the other is the dissociation energy of that cluster singly ionized. More sophisticated fission-barrier calculations have been performed by Landman and coworkers [76] for small $Na_N^{2+}$ and $K_N^{2+}$ clusters ($N \leqslant 12$) by combining DFT and molecular dynamics. These microscopic calculations have shown the possible existence of double-hump barriers, as in nuclear physics, and have also reproduced the emission of a charged trimer as the predominant fragmentation channel; unfortunately the method becomes very difficult to apply for large clusters.

It has been found convenient to express $F_m$ as the sum of two terms [77]:

$$F_m = \Delta H_f + B_m \ . \tag{49}$$

The first term is the heat fission,

$$\Delta H_f = E(X_{N-3}^+) + E(X_3^+) - E(X_N^{2+}) \tag{50}$$

and $B_m$ is the barrier for the opposite process of fusion of the two fragments to give the parent cluster, that is:

$$X_{N-3}^+ + X_3^+ \rightarrow X_N^{2+}. \tag{51}$$

Even in the jellium model, the calculation of $\Delta H_e$ and $F_m$ is a formidable task. Very often, either the parent cluster or the final products are not magic, so to obtain the energy of each isolated fragment one has to carry out a deformed jellium calculation, in which the positive background adopts shapes whose equilibrium deformation has to be determined in a self-consistent manner. Furthermore, a detailed description of the barrier requires solving the KS equations for a sequence of jellium shapes connecting the initial configuration, spherical or deformed, with the final one corresponding to two separated fragments, each at its equilibrium deformation. Several shape parametrizations of this kind exist [78].

The fission reaction $Na_{24}^{2+} \rightarrow Na_{21}^+ + Na_3^+$ has been studied [79], modelling the fissioning cluster by axially symmetric jellium shapes corresponding to two spheres smoothly joined by a portion of a third quadratic surface of revolution [80]. This family of shapes is characterised by three parameters: the asymmetry, $\Delta$, that is fundamental to the description of asymmetric fission, the distance parameter, $\rho$, which is proportional to the separation between the emerging fragments, and the "deck" parameter, $\lambda$, which takes into account the neck deformation.

Given a cluster configuration defined by a set of values of the jellium parameters $(\Delta, \rho, \lambda)$, the density of the valence electrons is calculated self-consistently by minimizing the total energy of the system. Fig. 9 shows the barriers obtained for two different jellium parameterizations, both of them characterised by $\Delta = 0.3134$. The first (dashed line) corresponds to the jellium shapes schematically shown at the top of the figure, where the cluster is forced to elongate up to $s = 18.3$ a.u. ($\rho = 1.175$) and scission occurs at $s = 23$ a.u. After that point, the barrier slowly tends from below to the classical Coulomb barrier (point-like coulombic repulsion between the fragments). The solid line in Fig. 9 corresponds

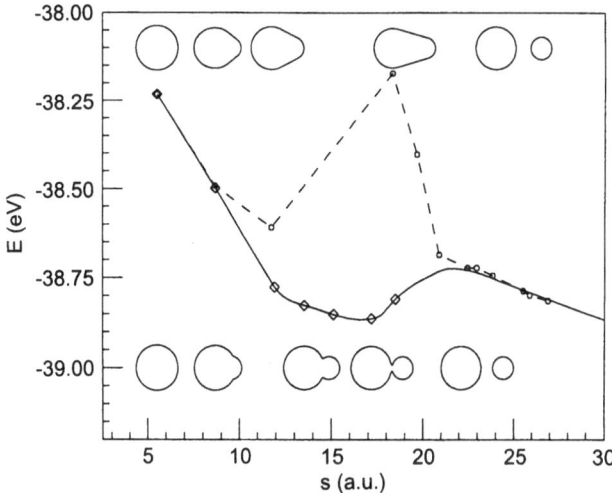

**Fig. 9.** Kohn–Sham total energy for the fission process $Na_{24}^{2+} \rightarrow Na_{21}^{+} + Na_{3}^{+}$ as a function of fragment separation for two different fission pathways. The dashed and continuous lines correspond to the jellium shapes schematically shown at the top and bottom of the figure, respectively [79].

to the fission pathway described by the jellium configurations shown at the bottom, in which the neck starts at $s = 6.1$ a.u. ($\rho = 0.35$). Other fission paths have also been studied; the result is that the solid line leads to the minimum barrier height, $F_m = 0.14$ eV. This is obtained as the difference between the energy at the maximum of the barrier ($s = 22$ a.u.) and the energy of the minimum at $s = 17$ a.u. This minimum is the ground state of the cluster, which is nonspherical because this cluster has the outermost shell only partially filled. The tendency of this cluster to asymmetric fission is already apparent in its ground state, which can be considered to be a supermolecular ion $Na_{21}^{+} - Na_{3}^{+}$. A similar result has been found for $K_{12}^{2+}$ in DFT molecular dynamics calculations [76].

A useful observation is that the saddle point corresponds to a configuration in which the emerging fragments are already separated and tied up by the electronic cloud. This is a general property, valid for asymmetric and symmetric fission channels of any fissioning cluster [77, 79]. As a consequence, if the goal is only to obtain the barrier height $F_m$ and not the full details of the barrier shape, one can start the calculation with a configuration of two tangent jellium spheres, and then increase their separation. This is the so called two-jellium-spheres model (TJSM) [81, 82], which has been employed to investigate the competition between fission and evaporation. Of course, an independent calculation has to be performed in order to obtain the ground state of the (spherical or deformed) parent cluster.

At very large separations, the interaction between the fragments is just the pure Coulomb repulsion between point charges. For separations near the touching configuration, the electronic densities of the fragments overlap, giving rise to a bonding charge similar to that in ordinary diatomic molecules. This bonding charge is responsible for the lowering of the repulsion energy below the pure coulomb repulsion; this lowering leads, in fact, to a maximum in the barrier [83].

Electronic shell effects are evident in the case discussed here, and affect many others. The separation of $F_m$ into the two terms of Eq. (49) is useful, because the shell effects are concentrated in $\Delta H_f$, as early studies had already hinted [75, 84]. For this reason calculation of $B_m$ has also been performed, using a simple version of DFT, namely an extended Thomas–Fermi (ETF) approximation, within the framework of the TJSM [81, 82]. The ETF functional differs from that in Eq. (11) in the kinetic energy, which is approximated by:

$$E_k^{ETF} = \frac{3}{10}(3\pi^2)^{2/3} \int n(r)^{5/3}\, d^3r + \frac{\beta}{8} \int \frac{(\nabla n(r))^2}{n(r)}\, d^3r \ . \tag{52}$$

The first term is the local Thomas–Fermi energy and the second gives the von Weisäcker quantum correction [11]. A value $\beta = 0.5$ for the coefficient in this term has been found convenient for jellium clusters [85]. The results obtained for $B_m$ are in good agreement with those from a full Kohn–Sham calculation. The ETF method is also useful to calculate the fission barrier $F_m$ for very large clusters [86], where the importance of shell effects is expected to decrease and a full KS calculation becomes tedious.

We have indicated above that the fusion barrier can be interpreted as the pointlike coulomb repulsion between "colliding" fragments, (with positive charges q − 1 and 1 respectively, plus a bonding (negative) contribution from the density overlap:

$$B(s) = \frac{q-1}{s} + V(s) \ . \tag{53}$$

For the charged-trimer emission, good agreement with the calculated fusion barriers of alkali metal clusters has been obtained by Garcias et al. [87], using the following parametrization of the bonding potential:

$$V(s) = -V_o R_o\, exp\left[-\alpha(s - R_o)\right] \tag{54}$$

in which $V_0 = 0.008/r_s \cdot r_s$ is the radius of the Wigner–Seitz cell in the metal, $R_0 = R_1 + R_2 = r_s\,[(N - 3)^{1/3} + 3^{1/3}]$ is the sum of the radii of the two fragments, and $\alpha = 0.2$. A qualitative justification for the form of $V(s)$ has been given using DFT [87]. Crucial for this justification is the exponential decay of the electron densities of the separated fragments (see Ref. [87] for details). A single value, $\alpha = 0.2$, describes the entire alkali group, but different values may be needed for other groups.

Combining the fusion-barrier calculated in this way with the heats of fission $\Delta H_f$ (see Eq. 50) obtained from a classical metallic drop model, Garcias et al. obtained the fission barrier heights $F_m$ for alkali metal clusters with charges

$q = 1\text{--}7$. The explicit expression for $\Delta H_f$ is

$$\Delta H_f = a_s [(N-3)^{2/3} + 3^{2/3} - N^{2/3}]$$
$$- \frac{1}{r_s}\left[\frac{q(4q-1)}{8N^{1/3}} - \frac{(q-1)(4q-5)}{8(N-3)^{1/3}} - \frac{3^{2/3}}{8}\right]. \tag{55}$$

This expression arises by adding in the metal-drop energy of Eq. (15), a term accounting for the coulomb energy of a charged cluster [70, 74], and neglecting the curvature term.

The heat of evaporation of a neutral monomer from the parent charged cluster $X_N^{q+}$ (see Eq. (47)) can also be calculated using the metallic drop model:

$$\Delta H_e = a_s [(N-1)^{2/3} + 1 - N^{2/3}] - \frac{q(4q-1)}{8r_s}\left[\frac{1}{N^{1/3}} - \frac{1}{(N-1)^{1/3}}\right]. \tag{56}$$

Comparison of $\Delta H_e$ and $F_m$ then leads to a prediction of the critical numbers for the observation of charged clusters. The results, given in Table 5, are in very good agreement with the experimental critical numbers, shown in parentheses in the Table, except for Na clusters with very high charge ($q = 6$ or $7$).

# 14 $d$-Electrons in Noble Metal Clusters

It was mentioned in Sect. 4 that electronic-shell effects appear in the mass abundance [10, 43], ionization potentials [88], and electron affinities [89] of noble metal clusters that are very similar to those observed for alkalis. These can be readily interpreted within the spherical jellium model if we treat the noble metal atoms as monovalent, that is, each atom contributes its external $s$-electron only. Even more, odd–even effects are also observed for small $N$ in the properties mentioned above, and have been explained by Penzar and Ekardt [32] within the context of the spheroidally deformed jellium model.

This provides information on the electronic structure near the top of the occupied orbitals of the cluster. The next question is how much deeper we can

Table 5. Calculated [87] and experimental (in parentheses [73]) appearance critical sizes for the observation of multiply charged alkali metal clusters as a function of charge $q$.

| $q$ | Li | Na | K | Rb | Cs |
|---|---|---|---|---|---|
| 2 | 24 (25 ± 1) | 26 (27 ± 1) | 24 (20 ± 1) | 24 (19 ± 1) | 23 (19 ± 1) |
| 3 | 56 | 63 (63 ± 1) | 59 (55 ± 1) | 59 (54 ± 1) | 57 (49 ± 1) |
| 4 | 103 | 117 (123 ± 2) | 110 (110 ± 5) | 109 (108 ± 3) | 105 (94 ± 1) |
| 5 | 164 | 185 (206 ± 4) | 173 | 172 | 165 (155 ± 2) |
| 6 | 240 | 268 (310 ± 10) | 249 | 247 | 236 (230 ± 5) |
| 7 | 330 | 366 (445 ± 10) | 337 | 335 | 319 (325 ± 10) |

probe down into the band structure of noble metal clusters. Smalley and coworkers [90, 91] have used Ultraviolet Electron Spectroscopy (UPS) to probe the $3d$ electrons of negative copper clusters ($Cu_N^-$) with $N$ up to 410 atoms. Probing the $d$-band requires *high* photon energies. These authors found a large peak, roughly 2 eV higher than the weak initial threshold, which moves smoothly with cluster size. For the small clusters, its position merges with the position of the $d$ levels of the copper atom. For the large clusters, the peak matches well with the sharp onset of the $3d$ band in the UPS of bulk copper. For all of these clusters, it seems safe to attribute this feature to the photodetachment of primarily $3d$-type electrons. Unlike the large size-dependent variations of the UPS threshold, which is associated to the $4s$ electrons, the $3d$ features shift monotonically with the cluster size, as is consistent with the different valence nature of these spectral features.

In the conventional band picture of solid noble metals, the valence band contains the localized $d$-electrons as well as the extended $s$-electrons, and $s$-$d$ mixing is substantial [9]. The picture of valence electrons is far from that of the free electrons in simple metals. It is, therefore, intriguing how well the shell model also works in noble metals.

Fujima and Yamaguchi [92] have performed selfconsistent DFT calculations for Cu clusters with sizes up to $Cu_{19}$ and a variety of model structures: $Cu_6$-octahedron, $Cu_8$-cube, $Cu_{12}$-icosahedron, $Cu_{13}$-icosahedron, $Cu_{13}$-cuboctahedron, $Cu_{15}$-rhombic dodecahedron. $Cu_{19}$-combination of cuboctahedron and octahedron, using the DV (discrete variational) $X\alpha$ method. An analysis of the molecular orbitals shows that these can be classified as two types. The first type is formed by molecular orbitals built from atomic $3d$ orbitals. These expand a narrow energy range of comparable width to that of the $d$-band of the solid, and do not mix with the second type of molecular orbitals, which are derived from atomic $4s$-$4p$ orbitals. The $3d$ charge is localized around atoms, whereas the $sp$ charge is extended over the whole cluster.

Next, Fujima and Yamaguchi tried to relate the results of their DV calculation to the shell model. If one disregards the molecular orbitals with $d$-character on the atoms, the sequence of the remaining molecular orbitals can be reproduced fairly well by considering a spherical model potential with a small unharmonic term (this is essentially the form of the effective potential one obtains in the spherical jellium model). However, if the cluster lacks a central atom, as in the case of the icosahedral structure of $Cu_{12}$, a 3-dimensional Gaussian potential barrier has to be added, to simulate the missing atom. The one-to-one correspondence between the energy levels of the DV method and the simple model potential leads to the conclusion that the shell model is applicable to Cu clusters. The $d$-band is located in energy between the $1s$ and $1p$ levels of the shell model for $3 \leqslant N \leqslant 8$ – more precisely between the molecular orbitals with overall symmetries comparable to those of the $1s$ and $1p$ levels – between the $1p$ and $1d$ levels for $9 \leqslant N \leqslant 18$, between the $1d$ and $2s$ levels for $19 \leqslant N \leqslant 20$, and so on.

# 15 The Spherically Averaged Pseudopotential (SAPS) Model

In spite of the success of the jellium model in explaining the electronic properties of simple metal clusters, in particular when deformations of the background away from the spherical symmetry are allowed, introduction of the granularity of the ions is a desirable next step for many purposes. A first-principles DFT approach is possible for small or medium size clusters, but solving the Kohn–Sham equations in this many-center problem, and especially the calculation of the equilibrium geometry, becomes a difficult computational problem for large clusters. With large clusters in mind, a method has been introduced [93] that goes one step beyond the usual jellium model, by employing pseudopotentials to describe the electron-ion interaction, although the method makes several drastic approximations concerning the net external pseudopotential. Those approximations make the problem of solving the Kohn–Sham equations tractable for large clusters.

Consider now a cluster where the ions are placed at positions $\{R_j\}_{j=1...N}$. If each ion is replaced by a local pseudopotential, $v_{ps}(|r - R_j|)$, then the total external potential seen by the valence electron cloud of the cluster is:

$$V_{ps}(r) = \sum_{j=1}^{N} v_{ps}(|r - R_j|) \ . \tag{57}$$

The usual calculation of the equilibrium geometry starts by assuming an initial geometry $\{R_j\}_{initial}$. The Kohn–Sham equations are then solved selfconsistently in the standard way to obtain the electron density and the total energy of the cluster. Since the initial geometry was chosen arbitrarily, small displacements of the ions from their initial positions can lower the total energy of the cluster. Another way of stating this is that the forces acting on the atoms are not zero for this initial configuration. An efficient way to continue is to displace each atom a small distance in the direction of the net force $F_j$ acting on it; this strategy is called steepest-descent relaxation. This process generates a new set of ionic positions $\{R_j\}$. The total energy of the cluster is calculated for the new geometry and the cycle is repeated again and again until all the forces vanish, that is, until the energy of the cluster is at a local minimum. This gives us one of the many possible isomers of the cluster. If we want to obtain the geometry corresponding to the absolute energy minimum, the entire process just described has to be repeated, starting with a new set $\{R_j\}_{initial}$ of "initial" ionic coordinates. Evidently after trying many initial configurations we have a better chance of finding the absolute equilibrium geometry, or at least one isomer with an energy close to it.

Other more sophisticated methods of calculating the equilibrium structure exist. One of the most effective is the technique of simulated annealing [94], which allows for surpassing potential energy barriers in the potential energy hypersurface, but it requires intensive computational effort.

Experience with the spherical jellium model suggests that, at least for clusters with a nearly spherical shape, one may simplify the process of solving the KS equations by replacing the external pseudopotential of eq. (57) by its spherical average about the cluster centre [93]:

$$V_{ps}(r) \to V^{SAPS}(r) \ . \tag{58}$$

In this way one arrives at the SAPS (spherically averaged pseudopotential) method. This simplification drastically reduces the computational effort since we now have a problem of interacting electrons that are moving in an external spherically symmetric potential well. Despite this simplification, the SAPS method goes a long way beyond the jellium model, since:

a) The SAPS potential is less smooth than the external potential of the spherical jellium model.
b) Although the ion-electron interaction only retains the radial part of the total pseudopotential, the ion-ion interaction $E_{ion\text{-}ion}$ is calculated for the true three-dimensional array of ions, that is:

$$E_{ion\text{-}ion} = \sum_{i \neq j} U(|R_i - R_j|) \ . \tag{59}$$

One can set limits of validity to the SAPS model. The cluster cannot be too small, because small clusters with open electronic shells deform away from spherical symmetry. On the other hand, very large clusters have the tendency to form planar surface facets. The intermediate cluster range between those two limits is well adapted to SAPS.

# 16 Results of the SAPS Model for Homoatomic Clusters

Figure 10 shows the calculated radial distribution of atoms in $Na_{25}$ and $Na_{30}$, taking as origin the cluster centre [93]. The empty-core pseudopotential [95] was used in the calculations. These results illustrate how the clusters are formed by shells of atoms. For clusters of small size, most atoms form a surface shell, and only a few atoms are in the inner region. The surface shell has a width of nearly one atomic unit. The evolution of the population of the outer and inner shells has been studied in detail for the case of $Cs_N$ [96]. Between $N = 7$ and $N = 18$ the centre is occupied by one single atom. The population of the inner region increases slowly after $N = 18$, forming an inner shell, leaving an empty hole at the centre of the cluster. At $N = 40$ one atom again occupies the center of the cluster and the configuration of a central atom plus two surrounding atomic shells persists until $N = 64$, when a third shell begins to grow in the inner region of the cluster. The restructuration mechanism for increasing $N$ is then: ($n$ shells) $\to$ ($n$ shells + 1 central atom) $\to$ ($n + 1$ shells).

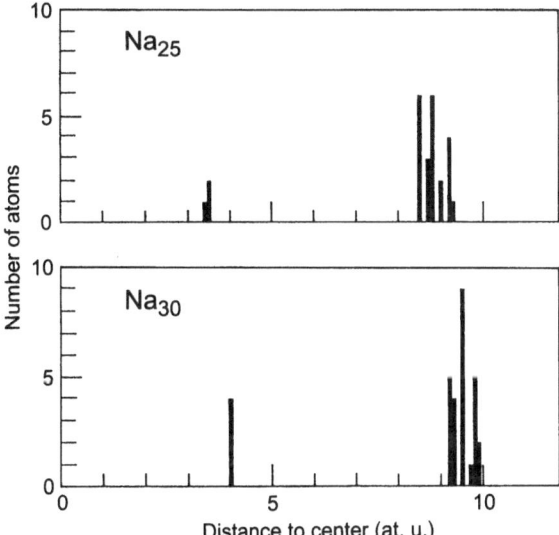

**Fig. 10.** Radial distribution of atoms in $Na_{25}$ and $Na_{30}$ calculated by the SAPS model [93].

In conclusion, a strong geometric reconstruction of the cluster occurs as it grows, at least in the case of simple metals. New atomic shells grow in the innermost region of the cluster when there is enough free space to accomodate first one single atom, and then additional atoms. There is a complementary view of this effect: The distance between the central atom and the surface atoms increases slowly between $Cs_7$ and $Cs_{18}$, since more and more atoms must be accomodated in the surface layer. For $Cs_{18}$ this distance has already become slightly larger than the nearest-neighbor distance, $d_{nn}^{bulk} = 9.893$ a.u., in bulk Cs. The SAPS calculation indicates that $Cs_{19}$ reconstructs its geometry in order to avoid interatomic distances larger than $d_{nn}^{bulk}$. This is achieved by placing the additional atom in the interior of the cluster rather than on the surface. For $Cs_{63}$ the situation is analogous. Now the central atom is surrounded by an atomic shell of 19 atoms with a mean radius nearly identical to that of $Cs_{18}$ again larger than $d_{nn}^{bulk}$. The next cluster, $Cs_{64}$, has two atoms in the innermost region, just like $Cs_{19}$.

A contraction of the cluster volume with respect to that of an equivalent piece of bulk metal has also been predicted [96]. The calculated cluster radius is smaller than the radius assumed in the spherical jellium model, where the volume is the same as that of an equivalent piece cut out of a macroscopic metal. This global contraction seems to be a general feature of small metallic clusters, and is well documented experimentally [97]. The volume contraction explains the discrepancies between experimentally determined static polarizabilities of small aluminium clusters and those obtained from jellium calculations [98]. The measured polarizabilities of $Al_N$ clusters with $N < 40$ are smaller than those predicted by a SJM calculation. The classical static polarizability (per atom) for a metallic sphere of radius $R$ is:

$$\alpha_{classical} = \frac{R^3}{N} \ . \tag{60}$$

For a jellium sphere the polarizability is enhanced, because the electronic charge density spills out beyond the jellium edge, and the polarizability can be written:

$$\alpha_{Jellium} = \frac{(R + \delta)^3}{N} \tag{61}$$

where $\delta$, which is related to the electronic spill-out, is small and approximately constant for all sizes. The fact that the jellium polarizabilities of small Al clusters are larger than the experimental ones indicates that $R$ is overestimated in this model. The SAPS calculation lowers the values of $R$ and leads to better polarizabilities [98].

At the end of Sect. 5 we mentioned the work of Lange and coworkers [27], noting how these authors modified the spherical jellium model in order to obtain the required sequence of magic numbers – reflected in the sudden drops of the ionization potential – for large clusters of alkali metals. Success was achieved by deforming the positive charge background to make the cluster denser in its inner part. SAPS calculations [99] have provided a microscopic interpretation of the model used by Lange and coworkers. An analysis of the interatomic distances in $Cs_N$ clusters with sizes up to $N = 80$ shows that the distribution of interatomic distances is not homogeneous, those in the inner region of the cluster being shorter than in the outer region. Additional calculations for Mg clusters [100] indicate the same effect. The inhomogeneous contraction of interatomic distances seems to be a general effect in simple metal clusters.

The SAPS model has been used to study the influence of cluster geometry on the photoabsorption spectrum calculated using TDLDA. Since only clusters with nearly spherical global shape can be treated by this method, we restrict the following discussion to this class of clusters. Photodepletion experiments for $Na_8$ [101, 102] show a single resonance peak at 2.53 eV whereas the SJM gives the plasmon peak at 2.92 eV. SAPS calculations, using a pseudopotential developed by Manninen [103], have been performed for two cluster geometries: $D_{4d}$ (square antiprism) and $T_d$ (tetracapped tetrahedron) [104]. These can be viewed as formed by one and two shells of atoms, respectively: the latter ($T_d$) is the ground state geometry, obtained from ab initio Configuration Interaction (CI) calculations [105]. For a fixed geometry, the position of the calculated plasmon peak depends on the interatomic distances. When these distances are taken from the CI calculations [105], the plasmon peaks are obtained at 2.52 eV ($T_d$) and 2.54 eV ($D_{4d}$), in very good agreement with experiment. It should be noticed that the CI interatomic distances are about 15% smaller than those measured in bulk sodium. The insensitivity of the peak position to the detailed structure of the cluster reveals the fact that the plasmon position is determined by its volume. Knowledge of the structure is, however, essential in order to describe the fragmentation of the plasmon, that arises from the presence of nearby particle-hole transitions. A two-peak structure has been observed in the photoabsorption spectrum of $Na_{20}$ [106]. The energies of the larger of these two peaks is 2.42 eV and that of the smaller one is 2.78 eV. The fragmentation of the surface plasmon is attributed to the proximity of a particle-hole transition. The SJM reproduces the two-peak structure, but the energies are larger, namely 2.72 eV and 2.98 eV respectively. A calculation for the geometry predicted by the SAPS model, which consists of a nearly spherical surface shell formed by 18 atoms enclosing two atoms inside it, leads to two peaks with energies closer to the experimental values [104].

For $Cs_8$, a fragmentation peak (at 1.48 eV) has been observed near the surface plasmon line at 1.55 eV [107]. Using the SJM, one obtains the surface plasmon at 1.79 eV, but not the fragmentation. The SAPS model predicts the square antiprism as the ground-state geometry of $Cs_8$. By properly choosing the cluster radius ($R = 9.09$ a.u.) and the core radius ($r_c = 4.6$ a.u.) for the Manninen pseudopotential, the SAPS model leads to a good fit to the experimental plasmon peak and to its fragmentation [104]. The radius adopted, $R = 9.09$ a.u., corresponds to interatomic distances 10% smaller than in bulk Cs. A good fit could not be obtained for tested geometries other than the square antiprism.

Non-local exchange-correlation effects have been considered in the study of $Na_{21}^+$ [108]. The measured spectrum shows a surface plasmon at $\sim 2.64$ eV [62]. The SAPS geometry is formed by a surface shell with eighteen atoms enclosing three others in the interior. The combination of geometrical (SAPS) corrections and nonlocal (WDA) effects leads to a peak position at 2.70 eV.

A promising extension of the SAPS model has been achieved by Schöne et al. [109]. These authors expand the external potential of Eq. (57) about the center of the cluster:

$$V_{ps}(r) = V^{SAPS}(r) + \sum_{i=1}^{\infty} \sum_{m=-l}^{l} V_{l,m}(r) Y_l^m \tag{62}$$

where the first term in the expansion is the SAPS potential. The second part of the pseudopotential was included perturbatively up to second order on top of a SAPS calculation. Selecting several isomeric geometries taken from ab initio molecular-dynamics DFT calculations [68], Schöne et al. obtained the same ground state geometry of $Na_8$ as in the ab initio DFT calculations. Incidentally, this geometry is different from the one obtained by the CI method [105].) The main problem with the perturbative approach of Schöne et al. is that in practice it can only be applied to rather symmetric clusters; so full geometrical optimization is not possible. Similar ideas, based on a perturbative introduction of geometrical effects beyond SAPS, have also been applied by Rubio et al [110] to $C_{60}$. Schöne et al. have also explored the post-SAPS effects on the collective electronic response.

In contrast, a simplified version of the SAPS model has been proposed by Spina and Brack [111]. Their main assumption is that all atoms in a given shell are at exactly the same distance from the cluster centre. An additional simplification is that the discrete point-like distribution of the ions in a shell is replaced by a uniform continuous distribution. In this way, the number of variational parameters corresponding to the ions is drastically reduced. The number of atoms in the shells, and the radii of these shells, are the variational parameters of the model. The results for the radii and populations of the shells agree well with those of the original SAPS, and make it possible to perform calculations for large clusters with very modest computational effort.

Lermé et al. [112] have applied the SAPS model to the study of shells and supershells in large clusters. For this purpose, they started with the simplified SAPS model of Spina and Brack [111]. The first aim was to investigate to what extent the granularity of the ionic background could modify the electronic shell structure of the SJM. Ab-initio DFT calculations [44, 67, 68] for small ($N < 25$) clusters and SAPS calculations for $N < 100$ [113] preserve the electronic shell structure of the SJM. The calculations of Lermé et al., for clusters with up to a few thousand electrons, show that in spite of the periodic distortions that modulate the effective potential, strong level-bunching occurs, characterized by the same bunching observed in the SJM. However, there are differences in the subshell structure. Lermé et al. [112] compared the results obtained with and without the simplification of Spina and Brack for the layer width, and they concluded that the subshell structure is sensitive to the details of the model as regards the ionic distribution. (The width of the ionic layers is influenced by the temperature.) They then turned to a study of the supershells. The introduction of pseudopotentials shifts the supershell nodes to lower electron numbers compared to the SJM, and the magnitude of the shift also depends on the width assumed for the atomic layers. However, the radial region close to the cluster center has no effect on the electronic shell structure, which is controlled only by the structure of the layers, near the surface. The precise parameterization of the atomic pseudopotential is also of a fundamental importance. The non-coulombic short range behavior of the pseudopotential results in an increase of the softness of the effective potential at the surface.

# 17 Formation of Shells of Atoms in Large Clusters

When the mass spectrum of large-$Na_N$ clusters is plotted versus $N^{1/3}$, the magic numbers appear at approximately equal intervals. However, the experiments of Martin et al. [14] show that the period of appearance of these features changes abruptly in the size region 1400–2000 atoms. The new periodicity, which is observed starting at $\sim 1500$ atoms and persists up to the largest clusters studied ($N \approx 22000$), is interpreted as reflecting the formation and the filling of shells of atoms. For small or medium size clusters, we know that the cluster shape changes every time an atom is added. However when the cluster size is large enough, changes in its global shape become more and more difficult and a new growth pattern emerges. It is believed that large clusters grow by adding shells of atoms to a rigid cluster core. The magic numbers observed in the experiments suggest that as the alkali metal clusters grow, they form closed-packed or nearly closed-packed polyhedra with icosahedral or cuboctahedral (fcc) shape. The total number of atoms $N_K$ in a cluster containing $K$ shells of atoms is [114],

$$N_K = \frac{2}{3}(10K^3 - 15K^2 + 11K - 3) \tag{63}$$

which are in very good agreement with the main features of the experimental mass spectra. Similar features are also observed for alkaline-earths [115, 116]. In this group an analysis of the secondary features of the mass spectrum is possible. This analysis gives information about the progressive formation of each shell, and indicates a "sub-shell" filling (faceting) process that is consistent with the icosahedral structure [115, 116]. It has not yet been possible to perform similar analysis for the alkali metals, but DFT calculations using the SAPS model give support to the icosahedral structure for these clusters as well [117].

The structure observed in the mass spectra obtained by Martin and coworkers [14, 115, 116] reflects size-dependent variations of the ionization potential. Although it is clear that the ionization potential will have a maximum value for a closed "electronic shell" and then drop, it is far from evident why a similar drop occurs after the closing of a "shell of atoms". SAPS calculations have been performed for model clusters with bcc structure and nearly spherical shape [118]. The clusters were modelled by starting with a central atom and adding a first atomic coordination shell of 8 atoms around the central atom, and then a second coordination shell of 6 atoms, etc. By proceeding in this way, the clusters modelled are nearly spherical and the SAPS approximation is more adequate. Clusters in the neighborhood of several atomic shell-closings ($N = 169, 331, 531, 941, 1243, 1459, 1807$ and $2085$) were explored by calculating the ionization potential IP as a function of size. IP suffers a drastic change after completion of a shell of atoms. In all cases there is a change in the slope of IP as a function of $N$, and a maximum is often observed there.

The structures detected in the experiments (cuboctahedral or icosahedral) have faceted surfaces and are less spherical than the model bcc clusters. However, we expect that the main result obtained for the model bcc clusters, namely the drastic change of IP after completion of a shell of atoms, can be extrapolated to more realistic geometries and to the relevant size range.

How can the transition from shells of electrons to shells of atoms be interpreted? Small sodium clusters are soft. There is no difficulty for the atoms to arrange themselves into a spherical conformation if this is demanded by the closing of an electronic shell or for the cluster to adopt deformed shapes in the case of open electronic shells. That is, small clusters behave like soft droplets, not necessarily liquid. When the size reaches about 1500 atoms, the electronic shell effects have become less intense and, consequently, changes in the global cluster shape become more difficult to attain. Under these circumstances, the formation of closed-packed symmetrical structures is more effective, and further growth takes place by condensation of atoms onto the surface of a rigid core, to form new shells of atoms.

The structure of bulk Na is body-centered-cubic. Consequently, the transition to the bulk structure has not yet occurred for $N \cong 20000$, in the Na clusters formed in Martin's experiments. Alonso et al. [119] have proposed that the reason why the bcc phase is not yet formed at these large sizes is that the screening cloud, $n^{scr}(r)$, around a $Na^+$ ion in a finite Na cluster depends so strongly on cluster size, that $n^{scr}(r)$ has not yet converged to its bulk limit even for clusters with ten thousand atoms. Since $n^{scr}(r)$ determines the effective interionic potential, which, in turn, determines the crystal structure of a metal [9], it is not surprising that much larger sizes appear to be required for the bcc structure of the bulk to develop. Further work along these lines is required in order to fully understand the screening of an ion in a finite cluster and its relation to concepts derived from bulk properties [120].

## 18 SAPS Model for Clusters with a Single Impurity

Here we are concerned with the question of how, embedding a highly reactive impurity, like oxygen, in an alkali-metal cluster changes the size-dependent electronic and structural properties of the host cluster. The case of Cs clusters with a single oxygen impurity ($Cs_NO$) has been studied in detail [99, 121, 122]. The only relevant comments about the SAPS calculation for this particular case are the following: a) The oxygen atom was placed at the cluster center. b) The inner electrons of the oxygen atom are also included in the calculation [99, 121]. The electronic configuration of the free oxygen atom is $(1s)^2 (2s)^2 (1p)^4$. The $1p$ shell becomes filled in the cluster by the valence electrons donated by the Cs atoms. Then the electronic shells of the cluster are filled in the following sequence: $3s$, $2p$, $1d$, $1f$, $4s$, $3p$, $1g$, $2d$,... This sequence results in closed-shell configurations when the number of Cs atoms ($N$) is 4, 10, 20, 34, 36, 42, 60,... The onset of the $4s$ shell is practically degenerate with $1f$ and the same $1g$ with $3p$. In conclusion, pronounced shell-closing effects only occur for $N = 10, 20, 36$ and 60. These shell closings are reflected in the calculated ionization potentials that display drops at these particular sizes. The main features of the experimental IP [122] agree well with the calculation: the predicted drops at $N = 10, 20, 36$ show up in the experiment, although the theoretical calculation exaggerates the oscillations of IP. In summary, the oxygen atom in $Cs_NO$ forms an anion $O^{2-}$ and the remaining $N - 2$ valence electrons of the cluster behave in the simplest way: that is, they give rise to shell effects. The experimental information [122] indicates that an analogous effect occurs for clusters with more than one oxygen atoms: that is, in clusters of composition $Cs_NO_x$ there are $N - 2x$ free-electrons.

A second aspect of interest concerns the structural effects induced by the presence of the oxygen impurity [99]. The strong ionic bonding between the oxygen and the Cs atoms produces a rearrangement of the inner part of the $Cs_N$ cluster. We can distinguish two cases: If the central site in $Cs_N$ is empty, the introduction of the oxygen atom produces only a small rearrangement of the innermost Cs shell: the radius and the population of this shell change only slightly. In contrast, when the central site in $Cs_N$ is occupied by a Cs atom, the structural rearrangement is drastic: a few Cs atoms, in most cases six, are split from the outer part to form a Cs shell directly surrounding the oxygen impurity. The radius of this shell is about 5 a.u. In contrast, the radius of the first coordination shell around a central Cs atom in a pure $Cs_N$ cluster is much larger. Part of the reason for this difference is that the atomic size of Cs is larger than that of oxygen, so Cs needs a larger hole to be accomodated. The strong ionic bonding also contributes. The population of the first coordination shell around the oxygen impurity is remarkably stable. With only a few exceptions, this shell is a group of six Cs atoms in octahedral arrangement around the O atom. As a consequence, the radius of this shell is also very stable. Both the sixfold coordination and the bond length in the $Cs_6O$ core agree with the corresponding properties of solid $Cs_2O$.

# 19 Mixing and Segregation Effects in Binary Alkali "Microalloys"

Experimental work on clusters containing similar amounts of two alkali-metal species is scarce. The main representative work has been done by Kappes et al [123]. These authors performed supersonic expansions of a mixture of lithium and sodium vapors. The most salient results are: a) mixed Na–Li clusters are produced; b) the magic numbers, revealed by the abundance in mass spectra, are the same as for the two pure species, that is, $N_e = 2, 8, 20, 40, \ldots$, where $N_e$ indicates the number of valence electrons in the cluster; c) there is an enrichment in the light element, Li, with respect to the initial composition of the mixed atomic vapors.

The fact that the magic numbers are the same as for pure Li or Na clusters is easily understood. If we think again in terms of the jellium-on-jellium model of Sect. 6, $n_+^0(Li) - n_+^0(Na) = 0.0029$ a.u., which is a very small number. This indicates that a jellium model, in which the jellium density is an average of the Li and Na densities, will predict well the magic numbers in the mixed clusters. However this "averaged jellium" model would be unable to say anything about the distribution of atoms in the cluster, or about mixing properties.

The SAPS model has been used for these purposes [124]. Steepest-descent relaxation, starting with a number of initial random geometries, was used to obtain the ground state geometries. The radial atomic distribution in $Na_9Li_9$ is given in Fig. 11 as an example. This cluster can be described as a single shell with radius $R \sim 7$ a.u., formed by seventeen atoms enclosing a Na atom at its centre. The picture is very similar to that in homonuclear alkali clusters [93]. Another feature can also be observed: the Na atoms are at a slightly larger distance from the cluster centre than the Li atoms. The phase diagram of bulk Na-Li alloys indicates complete immiscibility in the solid and a large miscibility gap in the liquid phase. This tendency appears to be less drastic in small clusters. The surface is responsible for reducing the demixing tendency: atoms on the surface have greater freedom to adjust their positions and to accomodate themselves in a convenient environment. This has been corroborated by calculating the heat of solution of a Li impurity in Na clusters:

$$\Delta E_{sol} = E(Na_N Li) - E(Na_N) - E(Li) . \tag{64}$$

The calculated values of $\Delta E_{sol}$ are negative for the sizes studied ($N \leqslant 20$). This explains why Na–Li clusters are formed in supersonic expansions. To study the enrichment of the clusters in Li the heat of the reaction

$$Na_n Li_m + Li \rightarrow Na_{n-1} Li_{m+1} + Na \tag{65}$$

was calculated. The values obtained are also negative, that is, substitution of a Na atom by a Li atom is favorable. During supersonic expansion, clusters are formed after many cycles involving atom aggregation and atom evaporation, so reaction (65) will be effective in promoting Li enrichment.

A phenomenon that sometimes occurs in bulk alloys is the preferential segregation of one of the components at the surface, that is, the enrichment of the outermost surface layers(s) in one of the components. The slight outward shift of the Na atoms relative to the Li atoms can be interpreted as a manifestation of this tendency. The difference between the Wigner–Seitz radii of Na and Li is $\Delta R_{WS}(Na–Li) = 0.72$ a.u. This difference increases for K–Na and Cs–Na, namely $\Delta R_{WS}(K–Na) = 0.87$ a.u. and $\Delta R_{WS}(Cs–Na) = 1.81$ a.u. Since the properties of alkali metals and alkali atoms vary smoothly with atomic number, we expect that the effects observed in Li–Na clusters will be enhanced for K–Na and Cs–Na clusters.

The radial atomic distribution in $K_{10}Na_{10}$ shows a drastic difference with respect to the Na–Li case: the Na and K atoms are well separated in "different" shells [125]. The Na shell has a radius, $R \approx 6$ a.u. and the K shell, or surface shell, has $R \approx 10$ a.u. The tendency of the heavier element to migrate to the cluster surface is common to Na–Li [124], K–Na [125] and Cs–Na [113] clusters, and is driven by the lower surface energy of the heavy element. This tendency is stronger in K–Na than in Na–Li. A difference between bulk Na–Li and K–Na alloys is that an ordered stoichiometric compound ($KNa_2$) forms in the second alloy but not in the first one. In fact, studying larger clusters like $K_{34}Na_{34}$, it is observed that not all the K atoms are segregated at the outer part of the cluster [125]; instead, there is an alternation of shells, which can perhaps be interpreted as a precursor of the ordering tendency in the bulk alloy. Evidently both much larger cluster, and the proper composition are required for this ordering tendency to develop fully.

Cs–Na clusters [113] show similar features to those already discussed for K–Na. The alternation of Na and Cs shells, and the presence of Cs at the surface, are consistent with the lower surface energy of Cs and with the existence of the bulk compound $CsNa_2$. Inspite of these structural effects, the electronic configuration in K–Na and Cs–Na clusters remains rather simple, and the electronic magic numbers are the same as for those of pure unmixed clusters.

Surface segregation also affects the collective electronic response properties. If we start with the ground state geometry of a mixed cluster – for instance, that in Fig. 11 – two types of structural change can be imagined: One is a simple exchange of the positions of several Li and Na atoms, and the other is a drastic change in the geometry. TDLDA calculations for K–Na clusters indicate that these two changes have different effects on the collective electronic response [125, 126]. The calculated photoabsorption cross section of $K_{20}Na_{20}$ is plotted in Fig. 12. The continuous curve corresponds to the ground state SAPS geometry. This can be viewed as an inner shell formed by eleven Na atoms plus a surface shell, split into two subshells: the outer surface-subshell containing all of the twenty K atoms and the inner surface-subshell formed by eight Na atoms. Finally, the cluster center is occupied by a Na atom. The segregation of K atoms at the surface is evident.

The calculated photoabsorption spectrum of this cluster shows a collective excitation with a peak at 2.12 eV. The tail of the resonance extends up to 3 eV, and concentrates a sizable amount of strength, due to particle-hole transitions that interact with the collective excitation and lead to its broadening. One of the most important particle-hole transitions is that from the HOMO level to the continuum; the energy of this ionization threshold is indicated by the arrow at $\sim 2.6$ eV. Similar TDLDA calculations have been performed for pure Na [104] and pure K [127] clusters. Comparing the positions of the collective resonances, it can be concluded that the position of the resonance in $K_{20}Na_{20}$ is closer to that in pure K clusters; thus, the surface, made of K atoms, controls the frequency of the collective resonance.

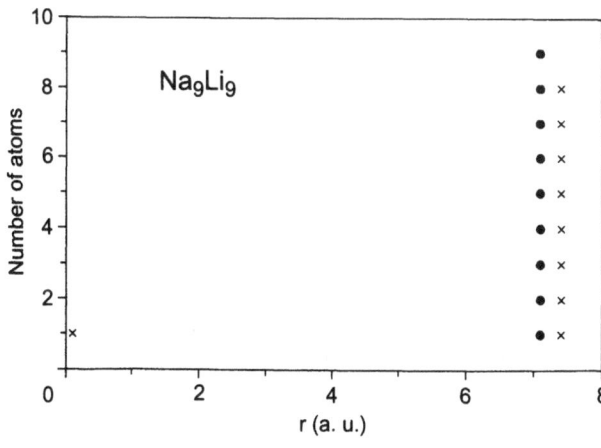

**Fig. 11.** Radial distribution of atoms in $Na_9Li_9$, calculated by the SAPS model [124]. Dots: Li, Crosses: Na.

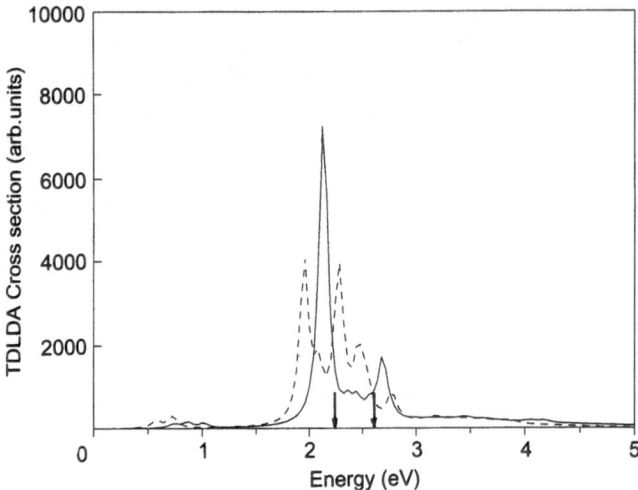

**Fig. 12.** Comparison of the calculated photoabsorption cross-sections for two isomers of the cluster $K_{20}Na_{20}$. The continuous curve corresponds to the ground state geometry. The left- and right arrows mark the ionization threshold ($-\varepsilon_{HOMO}$) of the isomer and ground state respectively [125].

If the positions of some Na and K atoms are simply exchanged, preserving the general architecture of the clusters, the shape of the calculated photoabsorption spectrum turns out to be very similar to the one of the cluster ground state, although shifted to slightly higher energies; that is, as the surface becomes enriched in Na the position of the collective resonance shifts smoothly towards the energies characteristic of Na clusters. The same effect has been found for other clusters studied: $K_nNa_n$, $n = 4, 5, 15, 26$ [126]. In all these cases the surface is rich in potassium and the plasmon peak occurs at an energy of 2.0–2.1 eV. Calculations on model structures in which the structure is conserved but the surface is enriched in sodium shift the peak towards higher energies while preserving the form of the spectrum.

A drastic geometrical change can be simulated by taking the geometry of one of the isomers that are usually found in the process of searching for the ground state. The dashed curve in the Fig. 12 corresponds to an isomer having a structure that – compared to that of the ground state of $K_{20}Na_{20}$ described above – has fewer atoms in the inner shell and on the outer-surface sub-shell, and more on the inner-surface sub-shell. The energy of this isomer is 0.12 eV per atom above that of the ground state. The prominent collective resonance is now fragmented, because the ionization threshold, indicated by the left arrow, interacts more strongly with the plasmon peak. In summary, the form of the photoabsorption cross section is sensitive to the cluster geometry and to the degree of segregation. This can be useful to ellucidate the structure of the clusters produced in the usual gas-phase experiments.

## 20 Ternary Clusters

The new features introduced by a third component have been studied on $Cs_nK_{10}Na_{10}$ clusters with varying $n$ [128]. In this case an efficient search for the lowest energy structure – or low lying relative minima close in energy to the ground state – necessarily requires the use of simulated annealing [94]. The most salient feature is the segregation of the Cs atoms at the surface. This is due to the lower surface energy of Cs with respect to K and Na. Another relevant observation concerns the location of the K atoms. It is convenient to begin by recalling that $K_{10}Na_{10}$ is a cluster with two shells: an internal shell formed by Na atoms and a surface shell formed by K atoms [125]. When Cs atoms are added to this cluster, these atoms prefer to sit on the surface. The geometrical information is collected in Fig. 13, which shows the positions of the different atomic shells, the position of each being identified by its mean radius. For small $n$, both the K and the Cs atoms can be viewed as forming the surface, although the mean radius of the Cs subshell is a little bit larger than the mean radius of the K subshell. This indicates a highly corrugated surface. As more and more Cs atoms are added to the cluster, more and more surface sites become occupied by Cs atoms and an increasing amount of K atoms lose direct access to the surface. In such a case, a migration of K atoms towards the inner part of the cluster is observed. Starting with $n = 21$, a few K atoms are still on or near the surface but

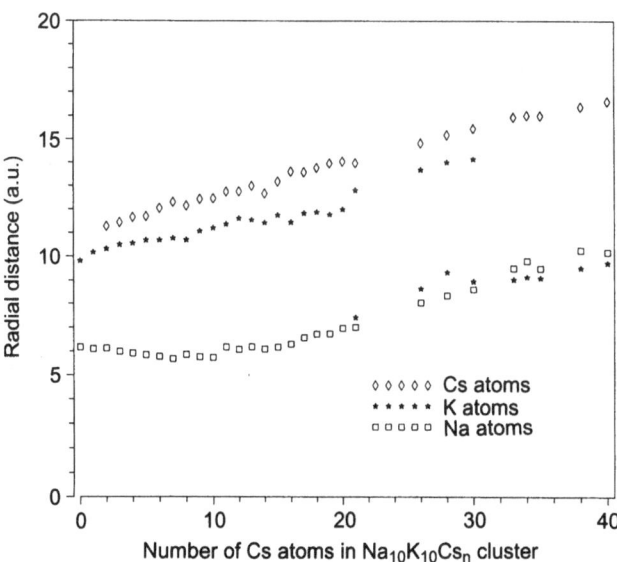

**Fig. 13.** Radial location of the different atomic shells in $Cs_nK_{10}Na_{10}$. The behavior of the K atoms is particularly interesting [128].

the majority have migrated to the inner region. The extreme situation occurs for $n \approx 33$, where all the K atoms have migrated to the interior of the cluster; the cluster can now be viewed as formed by a surface Cs-shell and an inner shell containing the K and Na atoms. Even more, for $n > 33$ this inner shell is formed by two well-defined subshells: the inner one contains the K atoms and the outer subshell the Na atoms: an striking inversion of the location of the K and Na subshells has occurred.

The peculiar behavior of the K atoms is a consequence of the interplay between two effects: the first is the surface effect, and the other is an electronegativity effect. The first one is responsible for the presence of K atoms at the surface of $K_{10}Na_{10}$, as well as for the formation of a Cs surface when this element is added of the $K_{10}Na_{10}$ core. However, as the K layer becomes increasingly covered by Cs atoms and the K atoms lose direct access to the surface, the electronegativity effect comes into play. As the electronegativity represents the power of an atom to attract electrons in a molecule or compound, electronic charge flows from regions of low electronegativity to regions of large electronegativity until the electronegativities become equalized. This charge flow has the effect of stabilizing the system, providing an ionic-type contribution to the binding. The electronegativities of Na, K and Cs are 2.70, 2.25 and 1.95 Volts, respectively, on Miedema's scale [9]. Evidently, a sequence of layers Cs–Na–K is more favorable, in order to maximize electronegativity differences, than a sequence Cs–K–Na since Cs–Na contacts have a larger electronegativity difference than Cs–K contacts. This, in our view, is the reason why the K atoms exchange their positions with the Na atoms.

## 21 Cylindrically Averaged Pseudopotential (CAPS) Model

Although very useful for certain classes of clusters, the SAPS model has its limitations for small or medium size clusters, because many of these are deformed. Nevertheless. deformed-jellium calculations indicate that most clusters still maintain axial symmetry (see Sect. 12) and that truly triaxial deformations are rare. These are good reasons to assume that the valence electron cloud is nearly axially symmetric, even if the ionic structure is fully three-dimensional. Taking up this idea, Montag and Reinhard [129] have developed the cylindrically averaged pseudopotential scheme (CAPS), which is the extension of the SAPS to axial symmetry. The essential approximation is to reduce the treatment of the electrons to axial symmetry and using cylindrical coordinates $(\rho, z)$. In this way, only the cylindrical average of the pseudopotential

$$V^{CAPS}(z, \rho) = \sum_{j=i}^{N} \bar{v}_{ps}(z, \rho; z_j, \rho_j) \tag{66}$$

$$\bar{v}_{ps}(z, \rho; z_j, \rho_j) = \frac{1}{2\pi} \int_0^{2\pi} v_{ps}(|\bar{r} - \bar{R}_j|) \, d\varphi \tag{67}$$

is seen by the electrons. A proper choice of the $z$ axis is critical for the success of the method. Montag and Reinhard considered the inertia tensor $\hat{I}$ of the ionic distribution and identified the $z$ axis with the principal axis of $\hat{I}$ whose momentum $I_i$ deviates most from the average momentum $\bar{I} = (I_1 + I_2 + I_3)/2$. The electrons see an axially symmetric potential and separate accordingly as:

$$\phi_{n\mu}(\rho, \varphi, z) = R_{n\mu}(\rho, z) e^{-i\mu\varphi} \ . \tag{68}$$

The Kohn–Sham equations were solved for sodium clusters on an axial coordinate space grid, and "softened" pseudopotentials were used.

The calculation of the ground-state geometry and electronic structure proceeds by an interlaced iteration of the Kohn–Sham equations and the ionic stationary conditions. The Kohn–Sham equations were solved by a damped gradient iteration method and the ionic configuration was iterated with a simulated annealing technique, using a Metropolis algorithm.

The results obtained show that the geometry of small Na clusters can be characterized in "slices" of ions with the same cylindrical coordinates $(\rho_j, z_j)$. The cluster can then be classified by the sequence $\{n_1, \ldots, n_k\}_{o/p}$ of $n_s$, the number of ions in a slice $s$, and the global shape – oblate (o) or prolate (p) – of the configuration. The ions on a slice are usually arranged on a ring, although in some cases an additional central ion appears in it. The structures predicted for several neutral and charged Na clusters are: $Na_2 - \{11\}_p$, $Na_3^+ - \{3\}_o$, $Na_4 - \{121\}_p$, $Na_6 - \{15\}_o$, $Na_7^+ - \{151\}_p$, $Na_8 - \{2222\}_p$, $Na_{10} - \{1441\}_p$. In all these cases there is agreement with the geometries predicted by ab initio DFT [44, 68] and configuration interaction [67, 130, 131] methods. For $Na_5^+$ the CAPS ground state geometry is $\{122\}_p$, and a low lying isomer $\{212\}_p$ was also found. Ab initio DFT and CI methods also predict these to be the two lowest isomers, although in the opposite order. To summarize, CAPS provides a reliable and efficient method to calculate the structure of metal clusters. CAPS can be very useful for studying the fission of doubly charged clusters, where the repulsion between the excess positive charges often leads to axial symmetry along the fission path.

# 22 Ab Initio Calculations

The description of simple metal clusters reviewed in this paper has been based on simple models: *spherical jellium model* (SJM), *deformed jellium model* (DJM), *spherically averaged pseudopotential model* (SAPS) and *cylindrically averaged pseudopotential model* (CAPS). These models allow us to perform calculations on medium and large size clusters, so that trends can be studied as a function of cluster size; numerous examples have been provided.

A detailed unconstrained consideration of the ionic structure requires a fully three-dimensional treatment of the electronic wavefunctions. Even if one replaces the effect of the atomic cores by pseudopotentials, the labor involved in solving the Kohn–Sham equations for a medium size cluster is substantial. If, in addition, we want to compute the ground state geometrical configuration of the cluster, the computational difficulties increase enormously. The pioneering work of Martins, Buttet and Car [44], starting with several reasonable candidate geometries and optimizing the structures with the steepest-descent method, allowed them to connect the ab-initio calculations with the SJM, and to provide reasons for the success of this model and its extension, the spheroidal jellium model. A crucial step has been provided by the introduction of the Car–Parrinello method [132] which combines DFT with molecular dynamics techniques to perform a simultaneous optimization of the electronic and nuclear (or ionic) degrees of freedom. At the present time, however, calculations for large clusters become cumbersome and the method is restricted to small clusters, usually containing no more than twenty atoms [68]. But this and other related ab initio molecular dynamics methods [133] are, no doubt, the most promising techniques for the treatment of medium- and large-size clusters in the near future.

*Acknowledgements*: This work has been supported by DGICYT (Grant PB92-0645).

# 23 References

1. Sinfelt JH (1983) Bimetallic catalysts. Discoveries, concepts and applications, Wiley, New York
2. Krätschmer W, Lamb LD, Fostiropoulos K, Huffmann DR (1990) Nature (London) 347: 354
3. Hohenberg P and Kohn W (1964) Phys Rev 136: B864
4. Kohn W and Sham LJ (1965) Phys Rev 140: A1133
5. Knight WD, Clemenger K, de Heer WA, Saunders WA, Chou MY and Cohen ML (1984) Phys Rev Lett 52: 2141
6. de Heer WA, Knight WD, Chou MY and Cohen ML (1987) Solid State Phys 40: 93
7. Bréchignac C, Cahuzac Ph, Carlier F, de Frutos M and Leygnier J (1990) J Chem Soc Faraday Trans 86: 2525
8. Bhaskar ND, Frueholz RP, Klimcak CM and Cook RA (1987) Phys Rev B36: 4418
9. Alonso JA and March NH (1989) Electrons in Metals and Alloys. Academic Press, London
10. Katakuse I, Ichihara T, Fujita Y, Matsuo T, Sakurai T and Matsuda H (1986) Int J Mass Spectrom Ion Proc 74: 33
11. Theory of the Inhomogeneous Electron Gas, Eds. Lundqvist S and March NH Plenum (1983) Press, New York
12. Ekardt W, (1984) Phys Rev B 29: 1558; Beck, DE (1984) Solid State Commun 49: 381
13. Bjørnholm S, Borggreen J, Echt O, Hansen K, Pedersen J and Rasmussen HD (1990) Phys Rev Lett 65: 1629
14. Martin TP, Bergmann T, Göhlich H and Lange T (1991) Z Phys D19: 25
15. Pedersen J, Bjørnholm S, Borggreen J, Hansen K, Martin TP and Rasmussen HD (1991) Nature 353: 733

16. Martin TP, Bjørnholm S, Borggreen J, Bréchignac C, Cahuzac Ph, Hansen K, and Pedersen J (1991) Chem Phys Lett 186: 53
17. Nishioka H, Hansen K and Mottelson BR (1990) Phys Rev B42: 9377
18. Genzken O (1991) Mod Phys Lett B7: 197; Genzken O and Brack M (1993) Phys Rev Lett 67: 3286
19. Strutinsky VM (1968) Nucl Phys A122: 1
20. Mañanes A, Membrado M, Pacheco AF, Sañudo J and Balbás LC (1994) Int J Quantum Chem 52: 767
21. Lang ND (1973) Solid State Phys 28: 225
22. Bréchignac C, Cahuzac Ph, de Frutos M, Roux J Ph, Bowen K (1992) In: Jena P, Khanna, Rao BK (eds) Physics and chemistry of finite sytems: from clusters to crystals. Kluwer, Boston, p. 369
23. Brack M, Genzken O and Hansen K (1991) Z Phys D19: 51
24. Pellarin M, Baguenard B, Bordas C, Broyer M, Lermé J and Vialle JL (1993) Phys Rev B48: 17645
25. Pellarin M, Baguenard B, Broyer M, Vialle JL and Pérez A (1993) J Chem Phys 98: 944
26. Baguenard B, Pellarin M, Bordas C, Lermé J, Vialle JL and Broyer M (1993) Chem Phys Lett 205: 13
27. Lange T, Göhlich H, Bergmann T and Martin TP (1991) Z Phys D19: 113
28. Kappes MM, Schär M, Yeretzian C, Heiz U, Vayloyan A and Schumacher E (1987) In: Physics and Chemistry of Small Clusters, NATO ASI Series B, Vol. 158, Eds Jena P, Rao BK and Khanna SN, Plenum, New York p. 263
29. Baladrón C and Alonso JA (1988) Physica B154: 73
30. Ekardt W and Penzar Z (1988) Phys Rev B38 4273
31. Clemenger K (1985) Phys Rev B32: 1359
32. Penzar Z and Ekardt W (1990) Z Phys D17: 69
33. Lang ND and Kohn W (1971) Phys Rev B3: 1215
34. Seidl M and Perdew JP (1994) Phys Rev B50: 5744
35. Perdew JP (1988) Phys Rev B37: 6175
36. Engel E and Perdew JP (1991) Phys Rev B43: 1331
37. Rubio A, Balbás LC and Alonso JA (1990) Physica B167: 19
38. Müller U, Schmidt-Ott A and Burstcher H (1988) Z Phys B73: 103
39. Bréchignac C, Cahuzac Ph, Carlier F and Leygnier J (1989) Phys Rev Lett 63: 1368
40. Kappes MM, Schar M, Radi P and Schumacher E (1986) J Chem Phys 84: 1863
41. de Heer WA (1993) Rev Mod Phys 65: 611
42. Ho J, Ervin KM and Lineberger WC (1990) J Chem Phys 93: 6987
43. Katakuse I, Ichihara T, Fujita Y, Matsuo T, Sakurai T and Matsuda H (1985) Int J Mass Spectrom Ion Proc 67: 229
44. Martins JL, Buttet J and Car R (1985) Phys Rev B31: 1804
45. Lindsay DM, Wang Y and George TF (1987) J Chem Phys 86: 3500
46. Kappes MM, Schär M, Röthlisberger U, Yeretzian C and Schumacher E (1900) Chem Phys Lett 143: 24
47. Kohl C, Montag B and Reinhard PG (1995) Z Phys D 35: 57
48. Manninen M, Mansikka-aho J, Nishioka H and Takahashi Y (1994) Z Phys D31: 259
49. Knight WD, Clemenger K, de Heer WA and Saunders W (1985) Phys Rev B31: 2539
50. de Heer WA, Milani P and Chatelain A (1989) Phys Rev Lett 63: 2834
51. Beck DE (1989) Phys Rev B30: 6935
52. Rubio A, Balbás LC and Alonso JA (1991) Z Phys D19: 93
53. Alonso JA and Girifalco LA (1978) Phys Rev B17: 3735
54. Rubio A, Balbás LC, Serra LI and Barranco M (1990) Phys Rev B42: 10950
55. Stott MJ and Zaremba E (1980) Phys Rev A21: 12
    Zangwill A and Soven P (1980) Phys Rev A21: 1561
56. de Heer WA, Selby K, Kresin V, Masui J, Vollmer M, Châtelain A and Knight WD (1987) Phys Rev Lett 59: 1805
57. Brack M (1993) Rev Mod Phys 65: 677
58. Pacheco JM and Ekardt W (1992) Ann Phys (Leipzig) 1, 255 (1992); Z Phys D24: 65
59. Rubio A, Balbás LC and Alonso JA (1992) Phys Rev B46: 4891
60. Alonso JA, Rubio A and Balbás LC (1994) Philos Mag B69 1037
61. Ekardt W and Penzar Z (1991) Phys Rev B43 1322

62. Brechignac C, Cahuzac P, Carlier F, de Frutos de and Leygnier J (1992) Chem Phys Lett 189: 28
63. Tiggesbäumker J, Köller L, Lutz HO and Meiwes KH-Broer (1992) Chem Phys Lett 190: 42
64. Borggreen J, Chowdhury P, Kebaili N, Lundsberg–Nielsen L, Lützenkirchen K, Nielsen MB, Pedersen J and Rasmussen HD: (1993) Phys Rev B48: 17507; Selby K, Vollmer M, Masui J, Kresin V, de Heer WA and Knight WD (1989) Phys Rev B40: 5417
65. Lauritsch G, Reinhard PG, Meyer J and Brack M (1991) Phys Letters A160: 179
66. Hill DL and Wheeler JA (1953) Phys Rev 89: 1102
67. Bonacic–Koutecky V, Fantucci P and Koutecky J (1991) Chem Rev 91: 1035
68. Röthlisberger U and Andreoni W (1991) J Chem Phys 94: 8129
69. Reinhard PG, Brack M and Genzken O (1990) Phys Rev A41: 5568
70. Bréchignac C, Cahuzac Ph, Carlier F and de Frutos M (1994) Phys Rev B49: 2825
71. Bréchignac C, Cahuzac Ph, Carlier F and de Frutos M (1994) Phys Rev Lett 64: 2893
72. Martin TP (1984) J Chem Phys 81: 4426
73. Martin TP, Näher U, Göhlich H and Lange T (1992) Chem Phys Lett 196: 113
    Näher U, Frank S, Malinowski N, Zimmermann U and Martin TP (1994) Z Phys D31: 191
74. Bréchignac C, Cahuzac Ph, Carlier F, Leygnier J and Sarfati A (1991) Phys Rev B44: 11386
75. Iñiguez MP, Alonso JA, Aller MA and Balbás LC (1986) Phys Rev B34: 2152
76. Barnett RN, Landman U and Rajagopal G (1991) Phys Rev Lett 67: 3058
    Bréchignac C, Carlier Ph, de Frutos M, Barnett RN and Landman U (1994) Phys Rev Lett 72: 1636
77. Garcias F, Mañanes A, López JM, Alonso JA and Barranco M (1995) Phys Rev B51: 1897
78. Hasse RW and Myers WD (1988) Geometrical relationships of macroscopic nuclear physics, Springer-Verlag, Berlin
79. Rigo A, Garcias F, Alonso JA, López JM, Barranco M, Mañanes A and Németh J, Surf . Rev and Letters (to be published)
80. Blocki J, J. de Physique (1984) C60–489; Blocki J, Planeta R, Brzychczyk J and Grotowski K (1992) Z Phys A341: 307
81. Garcias F, Alonso JA, López JM and Barranco M (1991) Phys Rev B43: 9459
82. López JM, Alonso JA, Garcias F and Barraco M (1992) Ann Physik (Leipzig) 1: 270
83. López JM, Alonso JA, March NH, Garcias F and Barranco M (1994) Phys Rev B49 5565
84. Rao BK, Jena P, Manninen M and Nieminen RM (1987) Phys Rev Lett 58: 1188
85. Serra L1, Garcias F, Barranco M, Navarro J, Balbás LC and Mañanes A (1989) Phys Rev B39: 8247
86. Alonso JA, Barranco M, Garcias F and López JM (1995) Comments At. Molec Physics 31: 415
87. Garcias F, Lombard RJ, Barranco M, Alonso JA and López JM (1995) Z Phys D33: 301
88. Knickelbein MB (1992) Chem Phys Lett 192
89. Ganteför G, Gausa M, Meiwes-Broer KH and Lutz HO (1990) J Chem Soc Faraday Trans 86: 2483
90. Pettiette CL, Yang SH, Craycraft MJ, Conceicao J, Laaksonen RT, Cheshnovsky O and Smalley RE (1988) J Chem Phys 88: 5377
91. Cheshnovsky O, Taylor KJ, Conceicao J and Smalley RE (1990) Phys Rev Lett 64: 1786
92. Fujima N and Yamaguchi T (1989) J Phys Soc Japan 58: 1334
93. Iñiguez MP, López MJ, Alonso JA and Soler JM (1989) Z Phys D11: 163
94. Kirkpatrick S, Gelatt CD and Vecci MP (1983) Science 220: 671
95. Ashcroft NW (1966) Phys Lett 23: 48
96. Lammers U, Borstel G, Mañanes A and Alonso JA (1990) Z Phys D17: 203
97. Solliard C and Flueli M (1985) Surf Sci 156: 487
98. Rubio A, Balbás LC and Alonso JA (1990) Solid State Commun 51: 139
99. Mañanes A, Alonso JA, Lammers U and Borstel G (1991) Phys Rev B44: 7273
100. Glossman MD, Iñiguez MP and Alonso JA (1992) Z Phys D22 541
101. Wang CRC, Pollack S, Cameron D and Kappes MM (1990) J Chem Phys 93: 3738
102. Selby K, Kresin V, Masui J, Vollmer M, de Heer WA, Sheidemann A and Knight WD (1991) Phys Rev B43: 4565
103. Manninen M (1986) Phys Rev B34: 6886
104. Rubio A, Balbás LC and Alonso JA (1992) Phys Rev B45: 13657
105. Bonacic–Koutecky V, Fantucci P and Koutecky J (1990) J Chem Phys 93: 3802
106. Pollack S, Wang CRC and Kappes MM (1991) J Chem Phys 94: 2496
107. Fallgreen H and Martin TP (1990) Chem Phys Lett 168: 233
108. Rubio A, Balbás LC and Alonso JA (1993) Z Phys D26: 284

109. Schöne WD, Ekardt W and Pacheco JM (1994) Phys Rev B50: 11079
110. Rubio A, Alonso JA, López JM and Stott MJ (1993) Physica B183: 247
111. Spina ME and Brack M (1990) Z Phys D17: 225
112. Lermé J, Pellarin M, Baguenard B, Bordas C, Vialle JL and Broyer M (1994) Phys Rev B50: 5558
113. Mañanes A, Iñiguez MP, López MJ and Alonso JA (1990) Phys Rev B42: 5000
114. Mackay AL (1962) Acta Crystall 15: 916
115. Martin TP, Bergmann T, Göhlich H and Lange T (1991) Chem Phys Lett 176: 343
116. Martin TP, Naher U, Bergmann T, Göhlich M and Lange T (1991) Chem Phys Lett 183: 119
117. Wang Q, Glossman MD, Iñiguez MP and Alonso JA (1994) Philos Mag B69: 1045
     Wang Q, Iñiguez MP and Alonso JA (1995) An Fis (Spain), 91: 29
118. Montejano JM, Iñiguez MP and Alonso JA (1995) Solid State Commun 94: 799
119. Alonso JA, González LE and Iñiguez MP (1995) Phys Chem Liq 29: 23
120. Sonoda K, Hoshino K and Watabe M (1995) J Phys Soc Japan 64: 540
121. Lammers U, Mañanes A, Borstel G and Alonso JA (1989) Solid State Commun 71: 591
122. Bergman T, Limberger H and Martin TP (1988) Phys Rev Lett 60: 1767
123. Kappes MM, Schär M and Schumacher E (1989) J Phys Chem 91: 658
124. López MJ, Iñiguez MP and Alonso JA (1990) Phys Rev B41: 5636
125. Bol A, Martin G, López JM and Alonso JA (1993) Z Phys D28: 311
126. Bol A, Alonso JA and López JM (1995) Int J Quantum Chem 56: 839
127. Rubio A and Serra L1 (1993) Z Phys D26: S118
128. Bol A, Alonso JA, López JM and Mañanes A (1994) Z Phys D30: 349
129. Montag B and Reinhard PG (1994) Phys Lett A193: 380
130. Bonacic–Koutecky V, Fantucci P and Koutecky J (1988) Phys Rev B37: 4369
131. Bonacic–Koutecky V, Boustani C, Guest M and Koutecky J (1988) J Chem Phys 89: 4861
132. Car R and Parrinello M (1985) Phys Rev Lett 55: 2471
133. Sung MW, Kawai R and Weare JH (1994) Phys Rev Lett 73: 3552

# Author Index Volumes 151–182

*The volume numbers are printed in italics*